To our wives — who have learned to expect conversation about oil with breakfast, lunch, tea and dinner

Preface

The importance of oil in the world fuel economy need hardly be stressed. It is a sobering thought that the end of the era of adequate oil resources is already in sight, and that the supply position can be expected only to worsen. The need to develop other sources of fuel and chemical feedstock is clear for all to see. Nevertheless, the reluctance of many people to accept the reality of this situation, coupled with the time required to develop the new sources, renders a severe world-wide shortage of fuel a very real possibility.

In this context the recovery of every barrel of oil which can be garnered at an economic price is of paramount importance; when we bear in mind that only about one-third of the original oil-in-place can be produced by present technology at an economic price we gain some idea of the opportunities for improvement. Conservative estimates suggest, for instance, that the target for enhanced oil recovery in the United Kingdom offshore sector is at least twice the country's annual gross national product, while for the United States the target may be as much as four times the G.N.P. of that country. There is in consequence a lively and growing interest in methods for maximising the aggregate yield of crude oil from underground reservoirs. Coincidentally with these developments in production practice and reservoir engineering has come spectacular progress in the biological sciences, above all in the related and overlapping areas of microbiology, biochemistry and genetics. The current level of understanding of biological processes in molecular terms, together with the techniques now becoming available for manipulating living organisms in a genetic and biochemical sense, encourages a reconsideration of using microorganisms directly, in the oil-bearing strata themselves, to assist in the production of crude oil. Already real increases in oil production have been obtained on a small scale with biological systems chosen on an empirical basis. The possibilities now before us for the construction of bacterial strains finely tuned for the tasks of mobilising oil in the reservoirs have prompted us to undertake this review of past experience and future possibilities.

It is important to take action and begin development work now; otherwise, we feel, it will be too late and a series of great opportunities will have been lost. Microbiology itself has advanced so rapidly in recent years, and continues to do so at such a rate, that the techniques which will be needed are already available in principle and there appear to be few hurdles to their modification and application for the specific purposes we outline in this book. There is still in operation a sufficiently large fraction of the world's oil-producing fields and wells so that even if the new technology takes a decade or more to perfect there will be ample opportunity to apply it with great advantage. It is in this spirit that we have undertaken our survey and hope that it will prompt the initiation of research towards the target of enhanced oil recovery while there is still time to reap the rewards.

V. MOSES
D. G. SPRINGHAM

Contents

CONTENTS

CHAPTER 1

Enhanced Oil Recovery: A Potential Role for Microorganisms

It is widely known that of the oil-in-place in underground reservoirs only a fraction, ranging from perhaps 10% to 50%, is recoverable by current technologies: the world-wide average yield of recovery is about 30%. That portion of oil which cannot at the moment be profitably extracted would clearly be well worth having; the development of technically satisfactory and economically viable ways of doing so would have profound industrial, economic and social effects on both producer and consumer countries.

The technical problems associated with the more commonly discussed enhanced recovery procedures are considerable: this is particularly true for deep reservoirs, and above all for those located offshore under difficult operating conditions, for which the costs of drilling and production are high, and the well spacing is therefore maximised. Most of the potential methodologies remain in the development and pilot stages and their application to difficult situations such as the North Sea fields seems unlikely for the near future.

The elaboration of a successful enhanced recovery procedure must go through a succession of developmental phases, starting with laboratory studies and proceeding through pilot trials to full field operation. Furthermore, the benefits of applying the procedure must be realised at a rate which ensures that the acquisition of an elevated production yield is not seriously offset by prolonged maintenance and operating costs during the period of enhanced recovery. The application of any of these techniques involves capital and operating expenditures and there will be a significant delay before production wells show an increased yield, particularly if enhancement methods are instituted when primary and secondary production is already well into decline. Indeed, in some procedures the wells are closed entirely for a specified period; costs, of course, continue to be incurred. Initial expenditures and investment before enhanced production ensues can thus be heavy. It then becomes particularly important that the enhanced recovery, when it does come on stream, should be at a sufficiently high production rate to be cost-effective.

Because of their uncertain technical effectiveness and doubtful economic benefits, enhanced recovery procedures (tertiary production) are normally put into effect only when primary production (e.g. water, gas cap or solution gas drives) and secondary recovery (e.g. pressure maintenance and waterflooding) are no longer sufficiently productive, or the amounts of produced water become too large. Tertiary methods may then be applied in the hope of obtaining additional yields. There may be advantages to be gained from applying them earlier than at the end of secondary production but because they are expensive, and their rate and magnitude of increased production not reliable, it is difficult to justify testing them while relatively inexpensive secondary methods remain effective. So a vicious circle develops; tertiary methods tend not to be used until beyond the point when they might have been most effective.

1.1. VALUE OF ENHANCED RECOVERY

There are both strategic and economic considerations influencing the decision to use enhanced recovery techniques. Until recently the pattern of world-wide oil exploration and exploitation permitted production levels to be increased in step with demand by operating in new fields. Many oil fields were themselves in industrialised user countries and were thus primary resources. And as far as Western countries were concerned, the bulk of their imported crude oil supplies came from countries which, if not under their direct political control, seemed favourably disposed towards them. Most of the important producer countries appeared to possess sufficient political stability to ensure that supplies of oil to the Western industrial countries would continue and grow for the foreseeable future.

That this was fundamentally an unsound assessment on the part of the general population and of at least some of those in power was first clearly demonstrated in 1973, and has continued to become clearer ever since. The Western countries are now faced with increasing uncertainty about both the availability and price of their future supplies of crude oil. The reasons may be summarised thus:

1. decreasing rate of discovery of new fields;
2. decreasing political stability and friendliness towards the West of some important producer areas;
3. increasing costs of prospecting (e.g. offshore and remote land areas);
4. increasing costs of production (e.g. offshore operations, laying of undersea pipelines, poorer reservoirs, etc.);
5. continuous rise in global demand.

The prognosis, now widely appreciated, is that sooner or later within the next few decades the number of new drilling sites will decline while the difficulties of exploration, drilling and production will increase, leading to curtailment of supply and inevitable increases in real prices. Under these conditions the incentive to maximise the return from investment will be heightened, and it will become ever more important to raise the total yield from existing wells and fields rather than simply to open new ones. Nowhere will this be more acute than under the difficult operating conditions offshore or in other unfavourable climatic or geographical conditions. These factors may thus be instrumental in prompting the more serious development of enhanced recovery procedures; it seems reasonable to conclude that the proportion of the total offtake obtained in this way will increase as time goes on. All possible enhanced recovery procedures are going to come under renewed scrutiny; among them those involving the use of microorganisms offer potentially the greatest measure of cost-effectiveness.

1.2. CONVENTIONAL PROCEDURES FOR OIL RECOVERY

The largely empirical methods widely used by oil companies are very successful at extracting about one-third of the oil-in-place with the minimum of expense. An appreciation of the ways in which a further portion of the oil can be recovered (enhanced oil recovery) requires a basic understanding of the well-tried conventional procedures. We give here an outline account of these procedures for the benefit of microbiologists and others who may not be familiar with them. A much fuller, non-technical account may be found in reference 1.

1.2.1. The Nature of Crude Oil

Crude oil deposits are believed to have been formed from plant and animal remains deposited in marine sediments. Over long periods of geological time, processes of bacterial decomposition followed by chemical change at high temperature and in the absence of air produced the extremely complex mixtures found today.

A precise description of the chemical composition of crude oil is not practicable because of the complexity of the material (crude oils may contain hundreds of thousands of components) and the fact that oils from different sources differ greatly in composition. The differences are indeed sufficiently great to permit the identification of crude oil spillages from tankers at sea by capillary column gas–liquid chromatography. However, useful generalisations can be made. Crude oils are predominantly organic and consist mainly of hydrocarbons. Sulphur compounds are present in small amounts (sulphur content varies from less than 0.1% to about 7% by weight), whilst nitrogen, oxygen and metal-containing compounds comprise (usually) trace amounts.

The major hydrocarbon classes are alkanes, cycloalkanes (naphthenes) and aromatics. Unsaturated hydrocarbons are very rare. The number of carbon atoms in different compounds varies from 1 to more than 50 and enormous numbers of isomers are present. Since there are more than 6×10^{13} isomers of an alkane with 40 carbon atoms,[1] the complexity of crude oil is easily accounted for. The smallest molecules are gaseous, the largest solid.

Two important physical properties of crude oils are specific gravity and viscosity. Most hydrocarbons are less dense than water, the specific gravities of crude oils range from 0·78 to 1·0. Specific gravity is frequently expressed in terms of API (American Petroleum Institute) gravity where:

$$\text{API gravity (degrees)} = \frac{141 \cdot 5}{\text{specific gravity at } 65°F} - 131 \cdot 5$$

Viscosity is of great importance during the recovery of oil since it determines the ease with which an oil will flow. Oil viscosities can vary over a very wide range: some light oils having viscosities comparable with that of water while some heavy oils have viscosities seven orders of magnitude higher at 0°C which makes extraction by conventional methods impossible. The viscosity of liquids decreases with increasing temperature and heavy oils may be made sufficiently mobile for extraction by increasing the temperature by 200 or 300 °C.

Gases are found in solution in the crude and frequently form a separate gas cap above the liquid zone. The major component is methane with smaller amounts of ethane, propane, butane, iso-butane and carbon dioxide. Hydrogen sulphide is present in some crude oils and is highly undesirable because of its corrosive and toxic properties.

1.2.2. Where Oil is Found

Oil deposits occur under the land and under the sea-bed at numerous sites widely distributed throughout the world. At one extreme oil-bearing rocks may outcrop at the land surface and at the other may be located at depths of several thousands of metres as in the North Sea, the bulk of known deposits occurring at depths between 600 and 3000 m (1970 and 9840 ft) below the surface.

Oil can migrate through porous rock and tends to be displaced upwards by water. It therefore collects in large quantities only in porous rock, usually in sandstone or limestone, where further upward migration is prevented by a cap layer of impervious rock such as clay or shale.

Gas is often found in association with the crude oil and, due to its lower density, normally forms a separate layer or gas cap above the oil. Water, often with a high salt content, and therefore known as brine, is always present in the oil-bearing layer (connate water). Frequently a separate layer of brine underlays the oil. A water layer contiguous with the oil is called an aquifer, and is often in continuous contact with the surface via layers of permeable rock. At any given locality oil-bearing strata may occur in several layers at different depths.

Porosity and permeability are two particularly important parameters of oil-bearing rocks. Porosity is that proportion of the rock which consists of spaces between the rock particles. It sets an upper limit to the quantity of oil which can be contained in a given volume of rock. Permeability is a measure of the ease with which a liquid can pass through a porous rock. It is measured in darcys: a piece of porous rock 1 cm long and 1 cm^2 in section through which a liquid of viscosity 1 cP moves at a rate of 1 ml/s under 1 atm pressure has a permeability of 1 darcy. The permeabilities of oil-bearing rocks vary from over 1 darcy to fractions of a millidarcy. In the latter case oils will flow only with difficulty.

Porosity and permeability are normally related but, for example, a rock with pores which were not interconnected could have a high porosity together with a very low or zero permeability.

1.2.3. How Oil Deposits are Located

Geologists can predict where oil is likely to occur from a knowledge of regional geology, local stratigraphy and geophysical data. A favourable location, for instance, would be an anticline, where a layer of impermeable rock covered a porous rock of a type known to bear oil in other localities. Unfortunately the geological predictions are no more than a guide. The only certain way of deciding whether or not oil is present is by exploratory drilling. This is an expensive process even on land where access is easy and the prospective oil-bearing stratum is shallow. In the North Sea where thousands of feet of rock underneath hundreds of feet of water have to be penetrated by the drill, and where weather conditions are extremely unfavourable, drilling a single well will cost millions of pounds. Having located a field, estimating its boundaries, thickness and other properties in order to permit optimum exploitation requires more exploratory holes to be drilled and the information obtained from them must be considered along with other geological and geophysical data.

1.2.4. Drilling Methods

The drilling of holes (wells) deep into the earth or into the sea bottom is one of the most characteristic operations of the oil industry. It is a highly-developed engineering operation in which a 10 000 m deep hole may be drilled almost as a routine.

A drilling system consists of a bit which is forced into the earth by rapid rotation on the end of a series of drill pipes, hollow rods, which are joined end to end. Drilling down may thus be continued to the desired depth. A tower structure (derrick) permits the bit and the series of drill pipes to be lowered into the earth or withdrawn again and houses the motors to rotate the drill pipes. Drilling is performed by rapidly rotating the drill pipes and the attached bit; in soft rocks speeds of up to 250 rpm may be used. Various types of bit are used according to the type of rock being encountered. The actual cutting surfaces may be faced with tungsten or diamond and commonly have three rotating toothed cutters. All or part of the weight of the string of drill pipes can be used to apply pressure to the bit to enhance penetration. Up to 40 tonnes may be applied to a 12 in (30 cm) diameter bit. Bits have a very short working life and the drill string frequently has to be pulled up to the surface to make a replacement.

A fluid, called a drilling mud, now carefully formulated with chemicals to suit the drilling conditions, is continuously circulated down the shaft and allowed to flow back up to the surface. This has several functions: to cool and lubricate the drill pipes and bit, to flush up the rock cuttings to the surface where they can be removed after examination, and to seal off any porous rock strata which are encountered by the bit.

For a very deep well drilling might start with a 24 in (60 cm) bit. After a few thousand feet of drilling the drill string will be pulled up to the surface and a cylindrical metal casing pushed down to line the hole. This is cemented in place in the rock formation and drilling recommenced with a drill of a smaller diameter. This procedure can be repeated two or three times giving a hole of decreasing diameter.

In offshore fields the drilling operation has to be performed from a complex and expensive drilling rig. To minimise the cost a whole series of holes may be drilled from a rig at a single fixed location. Thus in the Forties field in the North Sea, the entire field, some 30 square miles (7770 ha) in extent, is exploited from only four fixed platforms. Each platform has up to 27 wells which deviate from the vertical by as much as 55°. A well may thus reach as far as 9000 ft (275 m) from the platform at a depth of 8000–11 000 ft (2440–3350 m).

1.2.5. Primary Recovery

If a well is drilled into an oil-bearing layer oil may be forced into the well bottom by a number of mechanisms. These include the expansion of gas in the gas cap (gas cap drive), expansion of gas coming out of solution in the oil (solution gas drive) and the flow of water in from an aquifer (water drive). If the driving pressure is high enough, oil may be forced up to the surface where it may arrive with enough force to spout high in the air as a 'gusher' unless it is controlled. Alternatively, a pump, located at the bottom of the well, may be used to lift the oil to the surface. In a steeply dipping oil-bearing layer oil may drain down by gravity to a well at a lower level (gravity drainage). In a given reservoir one system will usually predominate although several mechanisms may contribute collectively to oil production.

The rate of oil movement to the well will depend on factors such as the pressure in the oil-bearing layer, the pressure at the bottom of the well, oil viscosity and rock permeability. Oil will tend to flow preferentially through those zones in the reservoir which have the highest permeability. Gas and water in varying amounts will be produced along with the oil.

As the oil flows from the zones of highest permeability it may be displaced by water from the aquifer which in turn flows into the well, reducing the proportion of oil in the mix arriving at the surface.

The production of oil by these primary mechanisms will eventually slow down owing to the decrease in solution gas or gas cap pressure or due to a slowing down of the water drive. The proportion of the total oil-in-place recoverable by primary mechanisms may range from 10% to 50%, averaging perhaps about 25%.

1.2.6. Secondary Recovery

When oil production by primary processes declines use may be made of the so-called secondary procedures. These involve pumping a fluid into the reservoir through one or more additional wells either to maintain reservoir pressure or to displace the oil directly.

Waterflooding is the most common method of secondary recovery. Typically, extra wells (injection wells) are drilled to penetrate the reservoir below the oil level and fresh or salt water, as available, is pumped into the underlying aquifer. Commonly four injection wells will be located around a production well (the 'five-spot' pattern) so that oil is displaced inwards from the injection points towards the production well.

The salt content of the injected water may be critical because clay particles in the rock may swell or contract if the salinity changes. Filtration to remove suspended solids which might block ('plug') the formation and deaeration to prevent corrosion and bacterial growth being enhanced by oxygen are frequently practised. Descaling agents, defoamers, flocculants and antibacterial agents may also be added.

The efficient use of secondary recovery procedures will produce an additional proportion of the oil-in-place, bringing the total to perhaps 30% or 35% on average. Although primary and secondary procedures were traditionally used in sequence there is no reason why, in favourable circumstances, they should not be used simultaneously, as is happening in some of the North Sea fields.

1.2.7. Residual Oil

As oil production by primary and secondary mechanisms continues an increasing proportion of water occurs mixed together with the oil produced. This results from water moving into the well from the aquifer and tending to flow preferentially along the easiest paths, thus bypassing large quantities of oil. Eventually oil will flow so slowly or the oil/water ratio will become so low that operation of the well becomes uneconomic. The point at which this occurs will clearly depend on economic factors including the price of oil and the expenses of operating the well. Oil fields in difficult locations, such as the North Sea and the North Slope of Alaska, where operating costs are extremely high will clearly become uneconomical much more quickly than those in easily accessible sites where operating expenses are low. Political factors, including the need for foreign exchange, and access to ensured supplies, may prolong the life of a well or field beyond the point at which strict economic considerations would dictate closure.

Bypassing of the oil by water flowing in from an aquifer or pumped in from an injection well may be visualised on several different scales. On the largest scale whole regions of a reservoir may be bypassed by the displacing water and the oil-in-place left untouched. This may occur because of discontinuities in the permeable rock which prevent the inflow of water or the outflow of oil or because other regions of the rock offer a much easier passage for the water due to higher permeability. In thick oil-bearing strata the water may find a path underneath the oil. As a result, instead of eventually displacing the oil along a more or less linear interface, the water penetrates into the oil-bearing layer in a highly irregular manner known as fingering. Once a region of oil is bypassed in this manner the flow of oil tends to slow down because water, which is usually much less viscous, flows preferentially along channels of higher permeability. As time goes on, more and more breakthrough of water occurs, leading to a decreasing oil/water ratio.

Comparable processes go on on a smaller scale. Since the linear rate of movement of a fluid in a capillary is proportional to the second power of the diameter, oil will tend to be displaced more rapidly from large rock pores than from small ones. This again will tend to lead to quantities of oil being cut off by the advancing flow of water. Once oil is isolated by water in a capillary the area of the oil/water interface must increase if more oil is to be released. From a knowledge of the likely values of the oil/water interfacial tension and the local fluid velocity it can be shown that such oil, once entrapped, will not be released unless either the water viscosity, or its rate of flow, are increased to improbably high values, or the interfacial tension is reduced by about three orders of magnitude by the introduction of a surfactant into the water-flood (section 1.3.1.3).

1.3. BRIEF SURVEY OF NON-BIOLOGICAL ENHANCED RECOVERY

We will not here discuss current enhanced recovery methods in great detail; recent reports are available which do so.[1-5] We are concerned only with considering methods having relevance for microbiological developments, thereby providing ideas in a non-biological context which might lead to useful biological approaches.

Enhanced recovery methods fall into three main categories: thermal methods, chemical flooding and miscible displacement. Thermal methods encompass the injection of hot fluids (water, steam or gas) into the reservoir, and promoting *in situ* combustion or the use of bottom-hole heaters powered by gas or electricity. The primary intention is to reduce the viscosity of the crude oil in the reservoir and promote its mobilisation. There is no possible biological counterpart for such heating methods, except perhaps the use of biological agencies to reduce the viscosity of oil in the reservoir strata by chemical modification. The attendant difficulties are briefly touched upon in section 5.2.2.2; at the present time such an approach does not appear likely to be successful in large-scale operations.

1.3.1. Chemical Flooding

This process depends upon the injection with the waterflood of specific chemicals, or combinations of chemicals, intended to influence the displacement efficiency of the waterflood itself. The chemical substances employed in these procedures have been selected for their efficacy in performing defined functions, availability and price. They certainly have not been chosen for their relevance to microbial biochemistry, that is, with any thought to how microorganisms might produce them. Indeed, it is clear that major components of current chemical flooding techniques are quite beyond the possibility of biological synthesis. However that may be, it is entirely feasible to consider biological counterparts of these chemical components in which the substances are different in detail but may share sufficient properties in common with those of non-biological origin to be of value. In one technique, that of polymer flooding, it is already the case that some of the polymers in use are produced by bacteria, although the generation of the product takes place in a conventional industrial installation and not, as we will ultimately discuss, *in situ* in the reservoir.

1.3.1.1.   Micellar Flooding

The relatively new technique of micellar fluid flooding requires the injection of a micellar slug, displaced with a mobility buffer, which is in turn driven by a conventional waterflood. The fluid comprising the micellar slug constitutes a micro-emulsion made by blending alcohol, petroleum sulphonate and brine; petroleum itself is sometimes an additional component. The mixture is formulated as far as possible to meet the conditions of individual reservoirs. In suitable combinations the emulsions may be oil-in-water or water-in-oil. The aim of the technique is for the slug to form a miscible phase between the oil in the reservoir and the water of the waterflood, thus driving a bank of displaced oil ahead of the waterflood. The use of a polymer-thickened mobility buffer, by increasing the viscosity of that buffer and hence raising the mobility ratio (i.e. the ratio of the viscosities of the displacing and the displaced fluids), promotes a piston-like displacement effect and reduces the tendencies towards bypassing and fingering.

The difficulties encountered in use are in principle similar to those for all chemical flooding techniques. The cohesion of the micellar slug is difficult to retain because of differential absorption of some components during passage through the reservoir strata; this and other physical factors lead to degradation and loss of stability both of the micellar slug and of the mobility buffer which follows it. The seriousness of the degradation depends on local conditions, but in general is exacerbated by high temperatures, low permeability and the presence of divalent metal cations. The longer the distance over which the maintenance of cohesion is required (i.e. the greater the well spacing) the greater the likelihood of degradation. If the chemical and physical properties of such a micellar flood could be produced continuously by biological agencies located at the oil/water interface it should be possible to avoid these limitations, provided the stability of the biological system itself is adequate. Of course, one cannot expect to be able to produce exactly the conditions of a current chemically-based micellar flood. A biological method would depend on the right combination of biogenic substances being synthesised by the biological system for the particular reservoir under consideration.

Micellar flooding is regarded as a promising technique, though its application to deep, offshore fields suffers from the disadvantages noted above and from the considerable expense involved in the manufacture, transport, storage and injection of the materials. Bleakley evaluated 177 enhanced recovery projects; while he regarded micellar flooding as the consensus choice, only four projects were active at the time and the results from them were not conclusive.[6]

1.3.1.2. Polymer Flooding

As an aid to increasing the mobility ratio, polymer flooding is employed either in association with micellar slugs, as discussed above, or as a water-diversion agent to promote a more favourable flow pattern of injected liquids. The latter method has value in flooding sections of the reservoir with low permeability, since injected fluids, seeking a path of least resistance, will tend to flow through regions of high permeability. This may leave considerable quantities of oil bypassed in the low permeability areas. Having penetrated the high permeability areas, the higher viscosity of the polymer solution tends to reduce the flow through those areas and thus to make penetration of the low permeability areas more likely. Time-setting polymers have also been proposed for this purpose, though their use would obviously become more difficult to control with increasing distance from the point of injection.

The limitations of polymer flooding may be quite severe. Shearing of the polymer molecules during blending, pumping and passage of the solution through small apertures and restricted channels reduces their effect of increasing viscosity. When injected into low permeability formations the polymers may so reduce fluid flow as to render further flooding impossible. Polysaccharide polymers may be degraded and serve as substrates for contaminating bacteria and permit sufficient growth to result in clumps of organisms, which then plug the formation.

Polymers currently in use include polyacrylamide and polysaccharides of bacterial origin, notably xanthans. The latter are long-chain compounds excreted into the medium by particular bacteria growing under specified conditions, and are normal commercial products.

As with micellar slugs, polymer slugs suffer from degradation which becomes more severe with the passage of time as the slug moves further from the injection point. The high cost of polymers and the large quantities required with wide well spacings are further problems in their use. Bleakley's survey included 20 polymer floods, of which 6 were profitable and 10 were not.[6] Again one might expect marked benefits if one were able inexpensively to generate polysaccharide polymers by bacterial action at their most effective location in the neighbourhood of the waterflood/crude oil interface. Continuous production of this sort would obviate degradation problems, maintain continuous effectiveness and avoid the cost of buying-in, transport, blending, etc. The operator would have greater confidence of continual polymer effects throughout the enhanced recovery period.

1.3.1.3.   Surfactant (Detergent) Flooding

This acts by reducing the interfacial tension between the oil and water phases, allowing the waterflood to detach oil from its adhesion to reservoir rock and sweep it forward on the flood to the producing well. However, similar problems to those with micellar and polymer flooding are encountered, particularly the adsorption of surfactant on to reservoir rock material and its loss from the flood. Again, the high cost of surfactant makes the economic advantages of flooding with this material very doubtful.

Once more one could envisage the benefits of microbial action if surfactant could be produced on a continuous basis at the water/oil interface. A continual loss of material by adsorption would be acceptable if replacement quantities were being produced fast enough by the microorganisms in place. Furthermore, the actual quantities required would be expected to be less than in surfactant flooding since the surface-active materials would be produced at their site of maximum effect.

Various microorganisms are able to produce appreciable amounts of surfactants although these are not for the most part similar to the ones already under trial for detergent flooding. Little or no formal evaluation of microbial surfactants as oil mobilising agents has been made. It is likely that they would be effective at some level of concentration, though whether this offers the possibility of an effective practical mobilisation system is not yet known.

1.3.2.   Miscible Fluid Displacement

These methods are those based on the injection of fluids which form a miscible phase on contact with the oil.

1.3.2.1.   Solvent Slug

The basis of this method is the injection of a slug of solvent (e.g. an alcohol, light hydrocarbon, liquid petroleum gas, carbon dioxide, etc.) which dissolves the crude oil and is in turn driven by a displacing fluid to sweep along the solvent towards the producing well. The solvent dissolves the oil and hence results in its mobilisation. The method is limited by the capacity of the solvent slug to dissolve oil. As the solvent becomes enriched in oil its capacity to effect further solution is diminished and it ultimately becomes saturated.

We see no feasible way in which the use of bacteria could be of value in solvent slug displacement.

1.3.2.2. Carbon Dioxide Flooding

This is a particular case of miscible fluid displacement making use of the special properties of this substance. Not only is carbon dioxide very soluble in crude oil under common reservoir conditions but on solution the volume of the crude oil increases and its viscosity decreases. Both changes assist the mobilisation of the oil.

Very large quantities of carbon dioxide are required: estimates are in the range of $4000–10\,000\,ft^3$ ($112–280\,m^3$) of carbon dioxide injected for every barrel of crude oil recovered.[7,8,9] We see no way in which microorganisms can generate such quantities of carbon dioxide *in situ*, and are thus unable to conceive of a biologically-based recovery based on a generalised mobilisation with a carbon dioxide flood. However, as we shall discuss in due course, it may be reasonable to consider the bacterial generation of relatively small quantities of gas, including carbon dioxide, which might be of value in expelling oil entrapped in dead ends and side paths into the main flood stream. This would be a specialised application of carbon dioxide flooding for a particular purpose.

1.4. CONCLUSIONS

The most feasible way to use microorganisms *in situ* in the reservoir to enhance crude oil recovery appears to be by some adaptation of chemical flooding.

The main limitations of chemical flooding are costs of materials, degradation of the slug and probably the need to inject quantities of material much larger than those ideally required to mobilise the crude. These drawbacks are so severe in the general run of chemical flooding operations as to render them economically non-viable.[4]

Quite apart from the development problems attendant on the elaboration of microbiological techniques, and recognising that some of those problems may in the event prove insoluble, it seems reasonable to us to conclude that a *technically* successful microbial technique may avoid the economic limitations of a chemical flood. If the microorganisms are able to synthesise the required chemical substances *in situ* from a portion of the oil already in place, most of the difficulties of the chemical flood may be surmountable. We note in particular that the use as a microbial substrate of oil which can otherwise not be recovered could be regarded as entailing no financial costs. The continuous production by microorganisms of the requisite chemicals, together with the propagative properties of living organisms, offers opportunities for the effectiveness of chemical displacement very different from those presented by conventional flooding with a slug of chemical.

This book will deal in some detail with the problems of using microorganisms in this way, together with an account of attempts which have already been made, or are in progress, in a number of countries, and some suggestions for future deployment of this general methodology; in addition to an earlier version of the present study,[10] a number of brief reviews have appeared in recent years.[11-14] We readily concede that a great deal of research and development will be required before there is hope of success, but we see in the present state of microbiology and its neighbouring fields of study a degree of understanding and manipulative skill which places the necessary developments within the range at least of the conceivable. At this stage of our report we seek primarily to set the scene for our later discussion, to describe briefly the potential relevance of microorganisms for enhanced oil recovery and to indicate which of the non-biological technologies already under discussion are the most appropriate for microbial exploitation.

REFERENCES

1. Stockil, P. A. (ed.) (1977). *Our Industry Petroleum*, 5th edition. British Petroleum Company Ltd, London.
2. Stewart, G. (1977). *Enhanced Oil Recovery*. Dept. of Petroleum Engineering, Heriot-Watt University, Edinburgh.
3. Johansen, R. T. (1979). *J. Rheology*, **23**, 167.
4. Groszek, A. J. (1977). *Symposium on Enhanced Oil Recovery by Displacement with Saline Solutions.* BP Educational Services, London.
5. Brown, J. (ed.) (1979). *European Symposium on Enhanced Oil Recovery, 1978.* Institute of Offshore Engineering, Heriot-Watt University, Edinburgh (several papers).
6. Bleakley, W. B. (1974). *Oil and Gas Journal*, **72**, 69.
7. Doscher, T. M. and Kuuskraa, V. A. (1979). In: *European Symposium on Enchanced Oil Recovery, 1978*, ed. J. Brown, p. 225. Institute of Offshore Engineering, Heriot-Watt University, Edinburgh.
8. Shah, R. P. and Wittmeyer, E. E. (1978) *Fourth Annual DOE Symposium: Enchanced Oil and Gas Recovery and Improved Drilling Methods*, paper C1. The Petroleum Publishing Company, Tulsa.
9. Meyer, R. F. (ed.) (1977). *Future Supply of Nature-made Petroleum and Gas*: International Congress Sponsored by the United Nations Institute for Training and Research (UNITAR). Pergamon.
10. Moses, V. and Springham, D. G. (1979). *Enchancement of Oil Recovery Using Microorganisms.* Opportunity Series No. 24. Industrial Liaison Bureau, The Hague.
11. Cowey, F. K. (1976). In: *The Genesis of Petroleum and Microbiological Means for its Recovery*, Birmingham, England, p. 57. Institute of Petroleum, London.
12. Bubela, B. (1978). *APEA Journal*, 161.
13. Malik, K. A. (April, 1979). *Process Biochemistry*, 4.
14. Forbes, A. D. (1981). In: *Hydrocarbons in Biotechnology*, Canterbury England. September 1979, eds. D. E. F. Harrison, I. J. Higgins and R. Watkinson. Institute of Petroleum, London.

CHAPTER 2

About Microbiology—And Its Relevance for Enhanced Oil Recovery

2.1. GENERAL CONSIDERATIONS

In the previous chapter we have summarised some of the reasons for employing enhanced recovery techniques and briefly discussed a number of methods which have been used or proposed as means of actually increasing the oil flow. Except perhaps for *in situ* combustion, all of these methods involve the injection of materials into the formation in the expectation of a beneficial response. A number of implications follow immediately. The injected materials (with the possible exception of water) have to be purchased, or capital must be invested for their collection (e.g. gases from neighbouring formations). They have to be transported to the sites where they are used and there injected under pressure into the formation. Storage facilities may be required at the wellhead. Thus, in no case is the employment of enhanced recovery techniques without economic cost and there remains considerable doubt, as expressed by a variety of authorities, as to whether there follows any net economic benefit from any of the existing technologies. Indeed, more crude may be produced, but the additional revenue seems generally to be less than the additional expenditure.

Much of the difficulty follows from the lack of control of the injected materials once they have entered the oil-bearing strata, coupled with the long lead times before, at best, beneficial responses can be expected. Even under conditions of 'huff and puff' operations, with injection into a production well and production from the same well started again at some later date, there is often a period of weeks or months when even the non-enhanced production rate is forfeit. Clearly this deficit must be recovered before any net benefit is derived. In a system employing separate injection and production wells, the time lag between the start of material injection and response of production will be influenced by well spacing, formation permeability, pressure differentials, etc. Even in relatively narrow well spacings, often so narrow as not really to be economic, the lag may be weeks or months long. With the wider well spacings more characteristic of offshore operations the delay may run into years. There follow, as a consequence, not only problems of front end loading in terms of investment and recovery but, equally importantly, doubt as to the stability of the enhancement procedures over long distances and for long periods of time in the formation. A good deal of evidence suggests unfavourable deterioration of surfactant or viscosifier slugs under these conditions.

Has microbiology anything to offer? Microorganisms, among their endless variety, have enormous capacities for chemical synthesis, a wide range of elaborate products and substances being built up from simple organic compounds and in many cases, indeed, from inorganic substances. Microbes, like all other living things, are self-propagating and self-renewing. They are also more or less delicate and succumb to hostile environments. But the possibility is clear: if a suitable microorganism (or a collection of different ones) could generate in the formation a substance or group of substances favourably affecting the recovery of crude and if, furthermore, this could be done over extended periods and distances without deterioration of the biological system and without great expense in the supply of nutrients, one could envisage the development of a technologically and economically satisfactory enhanced recovery system.

The self-renewal and propagation properties are potentially those most critically different from a conventional 'passive' injection methodology. If a slug of surfactant, for instance, is injected into a formation, its dispersion through the stratum depends almost entirely on the flow characteristics of the waterflood. It is not conceivable that the surfactant will do much to seek out the most effective location for its action seen from the operator's point of view. Furthermore, as the slug travels through the formation it will tend to degrade. The cohesion of the slug may deteriorate by fingering, material may be lost by adsorption onto rock particles, and chemical or biological influences may change the characteristics of the surfactant itself. These effects are hardly likely to reverse themselves spontaneously.

Consider, then, how a living system differs. Admittedly the general characteristics of the oil-bearing formation would have to be compatible with a satisfactory degree of survival and rate of growth of the chosen organism(s). But if some of the bacteria are lost in the wilderness where they are not effective they become of no further consequence. What is important is the arrival at the necessary sites of action of a sufficient number of organisms to provide an inoculum, or seed population, which will then grow to an effective level because the cells are in the right environment to do so (e.g. their foodstuff is present just there). It is even conceivable that living cells could be so chosen or constructed that they would from a reasonable distance seek out and move towards the appropriate location. Considerations like these add a totally new dimension to the possibilities of enhanced recovery which are worth thinking about.

2.2. TYPES OF MICROORGANISMS

What are microorganisms and what evidence have we for thinking that they may be useful?

It will be convenient first briefly to survey the various categories and to discard those which offer little potential benefit. Later we can consider in a little more detail the one group, the bacteria, which shows promise.

2.2.1. Desirable Properties
Microorganisms display an enormous variety of properties and even before we consider in detail just how they may be deployed

in enhanced oil recovery there are a few simple discriminatory criteria which are easy to apply. If an organism or a group of organisms is to be injected into a formation some, at least, of the following properties would be essential:

1. small size, to permit most ready penetration through rock strata;
2. resistance to high pressure since many reservoirs are deep;
3. maximum tolerance of the high temperatures prevailing in a large number of economically important reservoirs;
4. ability to withstand brines and sea water, since these are often present in reservoirs or used for waterflooding;
5. non-fastidious nutritional requirements, the simpler the better. The ability to live and thrive on the simple mineral salts already present (or cheap and easy to add) in the waterflood, plus the facility to use part of the crude oil *in situ* as a carbon and energy source, are highly desirable properties;
6. the capacity to use the foodstuffs anaerobically (i.e. in the absence of oxygen gas) since molecular oxygen cannot be provided in sufficient amounts down-hole to last perhaps for years on end;
7. a biochemical constitution commensurate with the production in adequate quantities of an effective agent to promote mobilisation of crude oil;
8. the absence of any unacceptable properties which might lead to plugging of the formation with a consequent fall in permeability, or the production of chemical substances causing deleterious changes in the oil, corrosion downstream, etc.

2.2.2.   Protozoa

One group of microbes comprises the tiny animals, made up of one or several cells. They offer little advantage for our purpose. Generally they tend to be too big, much too delicate, far too demanding in their nutrition and too dependent upon molecular oxygen to be worth considering for injection into reservoirs. In addition they probably would not produce much in the way of useful chemical products (except, perhaps, carbon dioxide) and it would be out of the question to try to provide a remotely acceptable physical and chemical environment for them. We can forget about the protozoa.

2.2.3. Algae

These are the plant equivalents of the protozoa. They are really also too large and characteristically have rigid cell walls which would limit their squeezing through narrow orifices; the protozoa are better at least in this respect. (Being mostly soft-bodied they are more deformable and could more readily get round awkward corners.) And the algae are probably rather less sensitive than the protozoa to environmental hazards and are less demanding nutritionally, except for one totally hopeless requirement: most of them, like other plants, need light. We can forget about algae, too.

2.2.4. Fungi

Among the most versatile of all groups from a chemical point of view, the fungi or moulds encompass species which between them will chemically convert almost anything to anything else. Many of them are quite good, too, at resisting all manner of hazardous environmental conditions. But most of them have rather large cells and suffer one enormous disadvantage: they grow in the form of filaments. Not only does this mean that as they grow they tend not to bud off new cells which can float away and come to rest somewhere else (although a very high proportion produce spores), but the filaments they produce will obstruct channels through the rock, impede the flow of oil or water and have an unfavourable effect on permeability. Some are known to cause blockages in fuel lines and tanks in aircraft.

One group of fungi, the yeasts, do not grow in filaments as such: they form single cells, or small clumps of cells. However, any aggregation whatsoever of cells would be a disadvantage as the aggregates would tend to clog at narrow apertures. It would be best if there were no clumping at all and, anyway, yeast cells are also rather large compared with those of bacteria and so would be less mobile through the formation.

There is another limitation which applies equally to all three groups so far discussed. The most effective way of using microbes for enhanced oil recovery will almost certainly involve a deliberate putting together of the right collection of properties within a particular cell type. Modern techniques exist for doing this, as we will discuss below. They have been developed primarily in bacteria and are only just now being applied to higher organisms like protozoa, algae and fungi. So the existing techniques are much more readily developed for special purposes with bacteria and this consideration is of itself of very great practical importance.

2.2.5. Viruses

Of all the groups of microorganisms these are the smallest and simplest, but they are active only within other living cells. Because they are so simple they have no conventional chemical means of converting foodstuffs, or even of making their own substance, and for those reasons are totally dependent on other organisms to perform those functions for them. Since they cannot live independently they cannot possibly be used as primary organisms for enhanced oil recovery.

However, the techniques we have just mentioned for building organisms to order involve particular viruses in the intermediary stages. They would thus have a role in the technology of enhanced oil recovery, but indirectly for the production of other organisms rather than directly for injection into the reservoir.

2.2.6. Bacteria

This is the group from which one may find potential candidates for use in oil recovery. There are very many types and species of bacteria, and they come in all sorts of shapes and sizes. Some cells are large, some very small; some grow in pairs, chains, or clumps, or even in filaments, while others grow as single entities; some can swim, others glide, while others again can only float passively. Among them there are types capable of growing on almost any organic source of carbon. Thus, there are forms which will attack some hydrocarbons and hydrocarbon derivatives found in crude oil, and will derive all their requirements of carbon and energy from such a source. There are also those growing at least partly on inorganic carbon, and between them they produce a prodigious range of products. Many bacteria grow in the absence of oxygen, and indeed some cannot survive in its presence. That is not to say that as wide a range of carbon sources can be used in the absence of oxygen as in its presence, but the range is wide nevertheless.

There exist bacteria able to grow at very high temperatures, some can withstand high pressures and some readily accept high concentrations of salt. Some, indeed, will tolerate all these simultaneously. And, not least, the range of genetic manipulations which can be performed with some bacterial types (and hopefully be extended to others) is already extensive and developing so fast that it becomes ever more feasible to consider actually designing and constructing bacteria to perform a specified collection of tasks. It is to the bacteria that we must look for living organisms to help with enhanced oil recovery.

The breadth of bacterial properties is so wide and so varied that from among their whole range one can readily imagine selecting a collection of functions and activities of value in enhancing oil recovery. One problem, however, is that the ideal organism, encompassing all the desired properties within itself, is unlikely already to exist in nature. Two possible ways may be used to overcome that limitation:

1. the use of a suitable collection of different microorganisms (a consortium) which between them will possess the desired characteristics, and which collectively will produce the desired response;
2. the construction by suitable genetic means of a single organism which *in itself* combines all the necessary properties. We discuss these alternative strategies in Chapter 5.

2.3. HOW MIGHT BACTERIA CONTRIBUTE TO ENHANCED OIL RECOVERY?

In non-biological approaches to enhanced oil recovery chemical substances are injected into the oil-bearing strata. Once injected no further control can be exerted over their distribution, longevity or effectiveness, so that decisions on how to proceed must rely on a prior analysis of reservoir characteristics. Only under very fortunate circumstances of shallow reservoirs and low drilling costs could one imagine drilling a series of wells solely to investigate the course of enhanced recovery procedures. In deep or offshore fields the problems are exacerbated; in particular, the wide well spacings essential for deep, offshore operations introduce great uncertainties into the progress and prospects for individual enhanced recovery attempts.

Let us suppose that a desirable bacterial product (or products) can be identified and that a suitable organism (or collection of organisms) which can make it can be put together in such a way as to be able to function under reservoir conditions, capable of movement through the formation and able to use some part of the crude as a prime source of nutrient. An appropriate concentration of the bacteria, grown at the wellhead, would be injected continuously with the waterflood. The bacteria would be dispersed, no doubt unevenly, at low concentrations, through the formation, in broad terms reaching those parts through which the waterflood flows. Many will find themselves nowhere near an oil/water interface, will be deprived of nutrition and hence will die out. Some, by chance or because they will be attracted towards hydrocarbons, will congregate at the oil/water interface and use some of the oil to grow and generate product which will mobilise the crude. This will be a continuous process: as the interface is swept forward by the waterflood, the bacteria will tend to follow it and continue the mobilisation effect. Some will fall by the wayside and be eliminated from further consideration. But as long as enough of them keep up to provide an inoculum from which new individuals may be generated, the system will be self-sustaining.

The critical biological features are clearly the ability to function appropriately in the particular environment of the reservoir in question, and a degree of stability, both biological and biochemical, which will permit the organisms to go on doing so for however long it takes for the enhanced recovery operation to be completed. By stability we imply that the properties of the system remain as they were at the start and deteriorate neither because the progeny of the original cells change in some undesirable way nor as a result of contamination with other forms of bacteria which may be able to survive better in the reservoir yet not produce the material necessary for enhancement of oil recovery. A wide understanding of bacterial physiology is thus an essential requirement for the design of a successful recovery system.

2.4. BACTERIAL METABOLISM

Like any other living cell, a bacterium is a highly complex and integrated chemical organisation. Again, in common with all highly organised structures, it has an inherent tendency to decay, lose its complexity and, as a result, to die. Maintenance of the organisation requires the continual expenditure of energy in the performance of (mainly chemical) work; if cells are deprived of a source of energy (i.e. are starved) they die once they have exhausted any reserve material they may possess.

2.4.1. Growth and Competition

When conditions are good and nutrition is plentiful bacteria grow. Individual cells increase in size, and eventually reach a point at which they divide into two daughter cells. If conditions remain good, the process enters a new cycle, the daughter cells increasing in size before they in turn divide to produce yet another generation. In any set of conditions we may suppose that the available resources will be competed for by the variety of cells present. All may be of the same species and hence more or less identical. Thus, given equal access to the resources, all will grow and divide at about the same rate so that the population, while increasing in numbers, will not change in quality. However, it may happen that a particular daughter cell is born with a defect which impairs its biochemical behaviour. As a result it will grow less quickly, divide less frequently and in the end its progeny will be swamped in the competition for scarce resources by its more efficient cousins. The defect line is said to be biologically 'unsuccessful', and will disappear from view. Alternatively, a cell may arise which in the local circumstances is more successful than the norm. It grows faster, divides more frequently and outgrows its competitors. If the conditions for its favourable competitive advantage persist for long enough the more successful strain will outgrow its competitors and become the predominant form.

Competition of this sort can take place between species as well as within the population of a single type. Hence, in designing a bacterial system for enhancing oil recovery it will be essential to ensure that the form selected for use is the most successful in the environment in which it is to operate. This means that should any variant arise, or should contamination with another type occur, the original form will outgrow all others and will remain quantitatively the most important, thus ensuring a stable oil recovery process.

2.4.2. Products of Metabolism

We can thus see that bacteria use their nutrients for the dual purpose of obtaining energy for chemical, mechanical and other forms of work, and for producing more of their own substance as they grow and divide. A typical bacterial cell contains thousands of different chemical substances yet may be growing on a single compound as a source of energy and carbon. It will be appreciated that an elaborate network of reactions is necessary to effect all the chemical changes: both the building up of molecules for bacterial substance, and the breakdown of foodstuffs into simpler forms.

2.4.3. Biological Energy Requirements

The need for energy is usually associated with oxidation. Potentially the hydrocarbons of crude oil are excellent substrates for bacterial growth for exactly the same reason that they are such good fuels for heat generation and transport: they yield a large amount of energy on oxidation. A major problem with the use of hydrocarbons as substrates is that in the reservoir the oxidising agent of choice, molecular oxygen, is absent. Alternative oxidising agents, such as nitrate, are conceivable but little work has so far been done to see whether bacteria can in fact use them for the oxidation of hydrocarbons in the absence of oxygen.

Another approach altogether with regard to nutrition is to inject the nutrient into the reservoir together with the bacteria.

Although this would imply a very severe limitation on the versatility of the method it is an approach which has had some measure of success in certain reservoir situations, and we discuss this in Chapter 4. It has the advantage that the nutrient injected can be of the type which is metabolised in the absence of oxygen but carries the penalty of having to provide the nutrient and to inject it.

2.4.4. Excretion of Products

Not all the products of nutrient utilisation are retained within the bacteria. Some substances are excreted as waste products; these are compounds of which the cells, in their particular metabolic state, are unable to make further use, and are discarded. It is among such materials that we may look for substances to aid us in oil recovery. Other compounds may be deliberately excreted by the cells in order to perform a necessary biochemical function in the immediate environment outside the cells. One such molecule might be an enzyme able to catalyse an attack on a hydrocarbon molecule in the crude oil, secreted in order to convert the hydrocarbon to a form more readily absorbed by the cells. It is from the potentially wide range of excreted and secreted compounds that we would seek a suitable collection to promote oil recovery and then attempt to combine their production within a viable living system.

2.5. REGULATION OF BACTERIAL METABOLISM

The ability of bacteria to catalyse biochemical reactions depends upon their content of enzymes (all of which are proteins) and coenzymes (small, 'accessory' molecules which are not proteins). So any factors which influence the activity or quantity of an enzyme will affect its net rate of catalysis. At any one time a bacterial cell will possess many hundreds of different enzymic activities; the coordination of all of these into a coherent biochemical organisation is achieved by a comprehensive network of interacting regulatory functions.

2.5.1. Passage of Material Across the Cell Membrane

Enzymes are large molecules usually resident within a cell. If they are to catalyse the formation of products from substrates present in the medium, the substrate(s) must have ready access to the enzyme(s). Most enzymes are inside cells although some are present on the outer surface, while others again are actually released into the medium and are quite free from the cells. We would expect that enzymes for metabolising hydrocarbons in crude oil would be on the surface or in the medium because it is not likely that the hydrocarbons would be able to penetrate fast enough through the membrane surrounding the cells to be used as food. However, in *Acinetobacter* it has been found that undegraded hexadecane is able to accumulate inside the cells and this may be a phenomenon which also applies to other species. Microbial cells live for the most part in an environment which is relatively hostile, in the sense that the medium surrounding the cells is very different in composition from the cytoplasm inside. They rely on the cell membrane to exert selective control over the passage of molecules between the cytoplasm and the medium. In some cases the selective control elements are themselves proteins, acting in a manner not unlike that of conventional enzymes and demonstrating a similar degree of specificity: such proteins are called 'permeases'. Alteration of the activity or quantity of such a permease would affect the nature and extent of chemical interaction between the cell and its environment.

The release of molecules from cells may also be a controlled process. Our supposition, therefore, that bacteria may be able to consume a portion of crude oil in the reservoir and produce substances active in oil recovery presupposes either that some hydrocarbons will enter the cell or that appropriate enzymes will leave it, and that the product molecules themselves will be released at a sufficiently rapid rate to avoid significant interruption of metabolism resulting from retention of product.

2.5.2. Changes in the Quantities of Enzymes

Very extensive studies during the last two or three decades have shed a great deal of light on the mechanisms which control the rates of synthesis of a number of microbial enzymes. While we cannot yet be sure whether all enzymes are subject to regulation in this way, it is clear that at least two main regulatory categories exist. In one, the enzyme is made only in response to the *presence* of a specific substance (usually the substrate) in the medium or in the cell. Such substances are called 'inducers', and enzymes in this category are said to be 'inducible'. In another main category, the enzyme is made only in response to the *absence* of a specific substance; the substance is then called a 'repressor', and the enzymes are 'repressible'. It is possible, by certain genetic manipulations, to isolate strains for which the mechanism controlling the rate of synthesis for a particular enzyme has been destroyed, or at least rendered non-functional: the enzyme is then made, often in very high yield, irrespective of the presence or absence of its specific inducer or repressor. Such enzymes are said to be 'constitutive'. It may be of considerable advantage to use such a genetic mutant provided:

1. that it is genetically stable, and does not revert to the parental form from which it arose; and
2. that it is physiologically healthy and will not be overgrown by possible competitors which may gain access to the same milieu.

A bacterial system for enhanced oil recovery must implicitly or explicitly be influenced by these considerations. It must therefore be chosen or constructed with a view to long-term genetic and biochemical stability under conditions (in the reservoir) where no subsequent modification or control is possible. A thorough understanding of the factors governing biological stability is an essential prerequisite for the design of a successful technique.

2.5.3. Changes in Enzyme Activity

Quite apart from modifying the rates at which enzyme molecules are made, regulatory mechanisms can also affect the rate at which molecules already extant can function. In the phenomenon of 'feedback inhibition' (or 'retro-inhibition') the final product of a biochemical pathway inhibits the *activity* of the first, or other early, enzyme in the pathway which has produced it. In 'feedback repression', applicable to the repressible enzymes discussed above, the final product of the pathway, or a derivative of it, inhibits the *formation* of some or all of the enzymes in the pathway. The total effect of these two factors is to restrict the synthesis of product molecules to the organism's own requirement. In an enhanced recovery procedure, however, we may well be interested in an organism producing for our purposes more of a product than it would need for its own purposes. If we succeed in achieving this by suitably altering the organism we must make sure that we do so in a way which precludes reversion to its more normal state of affairs when it is beyond our control.

2.6. ROLE OF INHERITANCE AND ENVIRONMENT

We have pointed out that the chemical activities of a bacterium depend sensitively on the content and activity of its enzymes. All enzymes are proteins, and their catalytic properties depend in turn upon their secondary and tertiary configurations (i.e. upon their three-dimensional structures); these are determined solely by their primary structures, the order of amino acid residues in the one or more polypeptide chains of which they are composed. The primary structure, for its part, is governed by the genetic information contained in the specific genes which code for particular enzyme proteins. Genetic information is encoded in the form of sequences of nucleotide residues in the polynucleotide chains making up the molecule of deoxyribonucleic acid (DNA) which constitutes the bacterial chromosome. The ability of the microbial cell to use substrates and produce products, which we have already seen depends on its enzymes, is thus primarily determined by its genetic inheritance.

This takes place at two levels. The most direct genetic involvement is the fact that the enzyme primary structure is specified by nucleotide sequence. A change in the latter may result in an altered enzyme which might be either more or less active catalytically than the original form from which it was derived. Such changes, described as 'mutations', are genetically stable and are inherited by the offspring of the mutant. In the long period of biological evolution since life originated we presume that an enormous number of mutational varieties occurred by chance, and that the ones surviving to the present day are those which have proved most effective in the biological context in which they arose. This means that in a process of microbial enhancement of oil recovery, which is likely to take place in conditions differing in many ways from natural microbial environments, mutants other than those historically selected in nature might be expected to be the most efficient and appropriate efforts will have to be made to obtain them. The problem of high temperature is an example of this, and one to which we will return later.

The second way in which genetic constitution affects the overall catalytic capacity of a cell derives from the fact that those control mechanisms which we discussed earlier are themselves determined genetically. This whole complex of genetic information, with which every living cell of whatever type is endowed, is called its 'genotype'. At any one time, as a result of the interplay of all the control factors, the cell will possess a particular complement of enzymes, etc., which represents less than its aggregate potential for their synthesis and accumulation. The *actual* composition of the cell, as distinct from the total *potential* of properties represented by its genotype, is called the 'phenotype'.

Because it is the phenotype that describes the real cell of the moment, the state actually engaged in chemical activities, this is the condition of primary interest when considering a possible role in oil recovery procedures. But it is clear that the phenotype results from the complex interactions between the immediate environment of a cell and its genetic constitution. It is the environment which largely determines which aspects of a cell's controlled activities shall be expressed; it is the genotype which sets the limits to those varieties of expression.

The conclusion for enhanced recovery procedures must be self-evident: to be of value an organism must possess the appropriate potential range of information and must respond in the right way to the total environment of the oil-bearing stratum so that its actual properties also are appropriate to the task in hand.

2.7. PHENOTYPIC AND GENOTYPIC STABILITY

In an earlier section mention was made of the inherent instability of biological organisation. It was shown that although an individual may, for a whole variety of reasons, fail to survive its place in the population will be taken by other individuals not so disadvantaged unless, of course, conditions are so bad that no members of the population can live in them. In that case that local environment may be invaded by other organisms which are differently endowed (genotype) and constructed (phenotype), in ways which permit them to thrive in conditions unsuitable for the first form. This is the concept of selection, one that is universally applicable in biology and its extension to many facets of human endeavour is simple enough to make.

If microbes are to be used to aid oil recovery the type(s) chosen must be genetically sufficiently stable for their desired properties to be expressed in the particular reservoir for which they are intended for an indefinite period. Put another way, not only must the chosen microbe do what is necessary but it should also be so well adapted to the environment in which it is to operate that no chance mutant or contaminant could successfully compete with it. It is not possible to ensure a lack of competition with absolute certainty. There are certain precautionary measures which might be taken but it seems rather improbable that they could be made absolutely reliable. For example, there may be a particular property present in the cells before their modification for enhanced recovery which it is desired to eliminate. A mutant may then be isolated in which the property is rendered non-functional because of a change in the nucleotide sequence of the relevant gene. Random changes in genes take place continuously, and there would be some probability that the changed gene might revert to its former state, leading to the reappearance of the unwanted property. The chances of this happening can be much reduced by introducing two, three or more changes into the gene because the chances of all the changes reverting *simultaneously* to the original state (which would be necessary for restoration of the property) is the product of the probabilities of each individual reversion taking place. The risk is even further reduced if a large portion of the gene can be removed altogether by excision; it is possible to do this in many cases with very stable consequences.

The invasion of the reservoir by contaminants of other species is another source of trouble to be avoided. Again, this is not something which can be totally prevented although stringent measures can be taken to eliminate unwanted species and admit only the ones permitted. Even if a few undesired individuals do slip through the precautionary measures they would fail to propagate to a significant extent if the chosen form was so much better adapted to the reservoir conditions that contaminants could not compete with it. Fortunately for this aspect of the matter, reservoir conditions are often so unfavourable to living organisms that the problem of finding bacteria capable of colonising them is likely to be greater than the problem of eliminating contaminants.

2.8. ENVIRONMENTAL LIMITS TO GROWTH

We have noted that oil reservoirs are unfavourable locations for microbial growth. The main hazards include high temperature, high pressure, unsuitable pH, high salinity, absence of molecular oxygen and lack of nutrients, together with physical factors such as formations so 'tight' that no free-living microorganisms could penetrate them.

Most of these limitations do not exclude the presence of living organisms at some level, although in the limit most of them could become so extreme that life would be quite impossible. Only the absence of molecular oxygen, among the factors listed above, would not limit life even if it were entirely excluded; however, the maximum range of biochemical activities would undoubtedly be restricted in its absence. Furthermore, not all of these unfavourable considerations would necessarily be present to the same extent in any one environment, and their interactive effects would inevitably be very variable.

Bacteria can in some cases live at remarkably high temperatures. They have been isolated from hot springs and other high temperature natural locations. There will be some upper limit beyond which no active life of any sort is possible but it is not yet clear what that limit is. A number of reports speak of certain bacteria growing well above 80°C. A few papers record growth close to 100°C and in one or two instances bacterial cells have been found to survive (though perhaps not to reproduce) at temperatures up to about 104°C. It seems feasible that with such examples before us a suitable high-temperature form may be obtained by a judicious search for it.

High pressure is less of a problem. In general bacteria often survive fairly well at pressures up to 300 atm above that in their normal growth environment. At pressures of 500 atm and above deleterious effects usually become well pronounced. However, anaerobic bacteria have been cultivated at pressures as high as 1400 atm (about 20 000 lb/in^2 (137 900 kN/m^2)) and have been shown to be catalytically active at 1800 atm (more than 26 000 lb/in^2 (179 260 kN/m^2)). In addition there is evidence that some bacteria resist high temperatures better when they are subjected to high pressure. For example, a strain isolated from an oil well core grew at 65°C under 1 atm and up to 85°C under 400–600 atm. Clearly, it is reasonable to expect to be able to find organisms growing at sufficiently high pressures and temperatures to be of potential value in a large number of reservoirs, while recognising that some are simply too deep and too hot.

With pH and salinity, too (the latter being of particular importance when flooding is performed with sea water), many bacteria are known which grow readily in quite high degrees of acidity, alkalinity and salt concentration. Once more there must be limits to growth but these appear wide enough not to present insuperable problems.

Concerning the nutritional limits to growth, we have already considered the desirability of the microbial population employed for enhanced recovery being able to subsist on part of the crude oil under anaerobic conditions. Bacteria also need supplies of mineral salts. Some of these would be available in sufficient quantities in the flood water or formation water itself. It is likely that two elements, nitrogen and phosphorus, would be deficient and require supplementation to levels of about 30–40 g/tonne of injected flood water.

2.9. GENETIC MANIPULATION: ISOLATION OF DESIRED BACTERIAL FORMS

It will be appreciated from what has already been discussed that investment made in understanding the biochemistry and physiology of bacterial action may be of critical importance in the production of strains to perform particular functions under defined reservoir conditions. Nevertheless, it may prove impossible to specify chemically exactly the nature of the changes an organism would need to undergo in order to be fitted most closely to the task in hand. Recourse may be had to the isolation on an empirical basis of strains with the required properties.

2.9.1. Isolation of Strains by Selection

We have already introduced the concept of organisms growing in competition with one another, from which it follows that judicious manipulation of the environment in which bacteria are growing can be used to select some forms and reject others. The basis of the selection is to permit the desired variety to multiply and form colonies while simultaneously preventing the undesired forms from doing so. By two or three sequential applications of this technique properly applied it is possible to isolate the desired strain totally free from all other sorts of living organism.

Starting with a strain from which it is hoped to isolate the mutant form required, a large number of cells are treated with a physical or chemical agent promoting genetic mutation. Mutagenesis is a highly damaging and random process. Its deployment customarily depends on starting with a large population of a particular organism, comprising tens or hundreds of millions of individual cells, which are treated in an appropriate manner with a mutagen. The resulting population consists of three categories of individual, the relative proportions of which will depend on the intensity of the mutagenic treatment: cells which are undamaged, cells altered so extensively that they are no longer viable, and those still viable but altered in one or more of very many possible ways. The problem is to select (i.e. 'isolate') from this mixed population pure clones with the desired phenotypic (and hence genetic) characteristics. Selection requires the designing of a growth situation in which only cells with the desired properties will multiply. When proper selectional procedures can indeed be designed to meet a particular situation the problem of strain isolation is simple, and the more selective the procedures (i.e. the greater the efficiency with which they can exclude unwanted forms), the simpler the isolation of the sought-after strain.

The frequency with which individuals with particular mutations arise in a mutagenised population may be very low, perhaps only 1 in 10^6 or 10^8 cells. Selection procedures need to be very discriminating to find the mutants. Suppose one wishes to find a strain capable of metabolising a hydrocarbon anaerobically: it may be sufficient to inoculate vast numbers of mutagenised cells into a growth vessel in which the hydrocarbon is the only source of carbon and energy, replace all the air in solution and in the gas phase with nitrogen or an inert gas, and incubate the vessel for a suitable period. To a first approximation only those forms able to use hydrocarbon anaerobically will grow. But there are possible pitfalls. The anaerobic users of hydrocarbon may indeed be the first to grow, but some of them may die and lyse (dissolve), thereby providing nutrient for other forms, eventually resulting in a mixed culture. Or again, anaerobic hydrocarbon users may be so rare, even after mutagenesis, that it is not possible to conduct the selection on a sufficiently large scale to ensure the presence of even one competent cell which could ultimately form a population. So no growth at all will take place.

Perhaps one wishes to isolate a form capable of living at a temperature somewhat higher than normal. A selection might now be attempted using a chemical milieu suitable for the parental strain before mutagenesis, but in which incubation is carried out at the higher temperature. That might produce the high-temperature form, although ability to grow at high temperature is unlikely to be determined in a simple genetic way. But say one wanted to find a form simultaneously able to grow at the elevated temperature and able to use hydrocarbon anaerobically. Could one select for both properties simultaneously? One could indeed try, but as the probability of finding the doubly mutated strain is the product of the probabilities for the two single mutations, a better tactic would probably be to isolate first one and then further modify that to get the second. Thus a whole series of changes would probably necessitate a whole series of mutageneses and selections.

2.9.2. Genetic Exchange Between Cells

It may be of benefit to consider combining properties which already exist in different organisms into a single strain. A number of processes are known among bacteria which make this feasible. Some bacteria undergo a sexual conjugation in which genetic material (DNA) is passed from a donor cell (the male) to a recipient cell (the female); different genes originating from both cells are combined in the female. In another technique, use is made of certain viruses which infect bacteria. They may grow and multiply within the infected cell, eventually killing the host and bursting its cell so that the new daughter viruses are liberated to the medium from whence they can infect further cells. During this process random pieces of bacterial DNA may be incorporated into the viral DNA. Thus, when reinfection occurs, some of the DNA from the original bacterium is transferred to the second. Under some conditions the infecting virus does not kill the host bacterium, but its DNA may become incorporated in a fairly stable manner into the new host, either by integration with the bacterial DNA itself, or by existing as a separate autonomous piece of DNA in the bacterial cell. Either way, part of the DNA from the original bacterium may be lodged in a new host in a more or less stable fashion. Yet further stages of transfer are possible and several different individual mechanisms are recognised. In summary, proper use of these viruses permits the construction of a more diverse range of genetic combinations than could be obtained just by mutagenesis and selection.

2.9.3. Genetic Engineering

Even with these techniques, however, limits are encountered. No virus may have yet been discovered to infect a bacterium of interest, or known viruses may not have the right properties. Some genetic combinations are so unlikely that no practical amount of searching has any hope of finding strains with the sought-after characteristics. The discovery of new categories of microbial enzymes in the last decade has opened up avenues of genetic manipulation undreamed of 10 or 15 years ago.

A bacterium's principal store of genetic information is located in the chromosome, a large circular loop of DNA, which is replicated more or less in step with cell division. Some, if not all, bacteria may contain additional smaller circular DNA structures, carrying defined genes, which replicate autonomously within the cell; these are called 'plasmids'. Plasmid DNA from a particular cell can be purified fairly easily by differential centrifugation in the right conditions, and separated from the main part of the cellular DNA in the chromosome. While some plasmids are normally transferred from males or females in a variant of the sexual process discussed above, it is possible in certain cases to reinsert DNA which has been extracted chemically from one cell into an appropriate recipient in which it may undergo recombination with the host chromosome; this is the process of 'transformation'.

Bacteria are able to protect themselves from the hazard of invasion by foreign DNA by the possession of two types of enzymes: modification methylases and restriction endonucleases. The former modify host DNA by methylating it at specific sites. The restriction nucleases cleave DNA not methylated in the manner specific for its species. It is thought that both enzymes may recognise the same portion of the DNA sequence. Thus, the endonuclease will not destroy a bacterium's own DNA because methylation will have taken place at the site where the endonuclease might attach, and hence will not be able to bind. Foreign DNA, not so methylated, will be cleaved.

Some restriction endonucleases cleave DNA at specific sites so that the breaks produced in the two strands are exactly opposite one another. Others cleave DNA only at sites at which a specific sequence of perhaps eight base pairs is present, the cleavage taking place in such a way that the cut in one strand is displaced from the cut in the other by four or five nucleotide residues. In one particular case, because of the symmetrical nature of the site cleaved, each of the cleaved ends terminates in –T–T–A–A. Now, if susceptible DNA is cleaved in this way, and the endonuclease removed or inactivated, the fragments will tend to associate, end to end, by virtue of base-pairing between the complementary ends, which are thus said to be 'sticky' (obliques represent limits of recognition sites; small arrows indicate cleavage points):

$$
X \quad \begin{array}{l} 5' \ldots \text{A}/\text{T–G}^{\downarrow}\text{A–A–T–T–C–T}/\text{A} \ldots 3' \\ 3' \ldots \text{T}/\text{A–C–T–T–A–A}_{\uparrow}\text{G–A}/\text{T} \ldots 5' \end{array} \quad Y
$$

cleavage

$$
X \quad \begin{array}{l} 5' \ldots \text{A}/\text{T–G} \quad \text{A–A–T–T–C–T}/\text{A} \ldots 3' \\ 3' \ldots \text{T}/\text{A–C–T–T–A–A} \quad \text{G–A}/\text{T} \ldots 5' \end{array} \quad Y
$$

separation

$$
X \quad \begin{array}{l} 5' \ldots \text{A}/\text{T–G} \quad\quad\quad \text{A–A–T–T–C–T}/\text{A} \ldots 3' \\ 3' \ldots \text{T}/\text{A–C–T–T–A–A} \quad\quad\quad \text{G–A}/\text{T} \ldots 5' \end{array} \quad Y
$$

reassociation between random fragments

$$
X \quad \begin{array}{l} 5' \ldots \text{A}/\text{T–G} \quad \text{A–A–T–T–C–T}/\text{A} \ldots 3' \\ 3' \ldots \text{T}/\text{A–C–T–T–A–A} \quad \text{G–A}/\text{T} \ldots 5' \end{array} \quad Z
$$

Finally, the cleaved bonds in the reassociated fragments may be reformed under the action of the enzyme DNA ligase ($\downarrow\uparrow$):

$$
X \quad \begin{array}{l} 5' \ldots \text{A}/\text{T–G}^{\downarrow}\text{A–A–T–T–C–T}/\text{A} \ldots 3' \\ 3' \ldots \text{T}/\text{A–C–T–T–A–A}_{\uparrow}\text{G–A}/\text{T} \ldots 5' \end{array} \quad Z
$$

In summary, a piece of DNA X ... Y is cleaved to X ... and ... Y which may separate; recombination takes place with other cleaved DNA containing ... Z to give X ... Z, and the bond is reformed with the ligase. The newly reconstructed plasmid is inserted into a bacterial host by whichever technique is most appropriate with the particular plasmid and host concerned. Clones in which insertion has been successful are then isolated by selection or screening as necessary.

2.10. CONCLUSIONS

This battery of genetic manipulative techniques offers enormous (though not literally endless) opportunity to construct bacteria to perform identified tasks in specified environments. At the present time general concepts and methods are already available. Each individual system will, however, differ in detail and one must expect to have to perform the special adaptation of those general methods to fit them to different organisms. Lengthy though these developments would be, there seems no reason for believing that they would not in principle be possible, though the nature of the experimental hurdles and ultimate limits to manipulation can for the moment only be guessed at.

GENERAL SUPPLEMENTARY READING

Very many textbooks dealing with the topics discussed in this chapter are available. Rather than list large numbers of them, we note below one good reference work in each of the areas of microbiology, biochemistry, bacterial genetics and genetic engineering:

Stanier, R. Y. Adelberg, E. A. and Ingraham, J. L. (1977). *General Microbiology*, 4th edition. Prentice-Hall, Inc., Englewood Cliffs, N. J.; Macmillan Press Ltd, London.

Lehninger, A. L. (1975). *Biochemistry*, 2nd edition. Worth Publishers Inc., New York.

Goodenough, U. (1978). *Genetics*, 2nd edition. Holt, Rinehart & Winston, Inc., New York.

Old, R. W. and Primrose, S. B. (1980). *Principles of Gene Manipulation* Blackwell Scientific Publications, London.

CHAPTER 3

Reservoir Microbiology

3.1 PRESENCE AND GROWTH OF BACTERIA IN RESERVOIRS

The success of enhanced recovery operations employing bacteria as envisaged here depends on the ability of the bacteria to survive and grow in the reservoir itself and it is pertinent to consider whether such survival and growth are possible. We shall consider both direct evidence, based on isolation of bacteria from oil, rock and produced waters, and indirect evidence based on changes in oil properties.

3.1.1. The Effects of Injection of Bacteria into Reservoirs

Since the 1950's there have been numerous field trials of enhanced recovery techniques involving the injection of bacteria into formations. These have been performed in the USA, Czechoslovakia, Poland, Hungary, Romania and the USSR (see Chapter 4) and although the results so far have not been of great economic significance, in the majority of cases there have been clearly measurable effects. These have included increases in oil production, increases in gas production, changes in oil viscosity, changes in gas composition and changes in water pH. These effects must have resulted from bacterial activity within the formation and, since in most cases only small quantities of bacterial inoculum were injected, bacterial growth must have occurred in the reservoirs.

3.1.2. Bacteria in Oil and Formation Rock

Several authors have reported the presence of bacteria in samples of oil and formation rock from the USA, the USSR and Germany (see, for example, references 1–6). Since it is normally impossible to take aseptic samples, and since the presence of large numbers of bacteria in drilling muds is well known, [1,7] there is room for doubt as to whether the bacteria originated from the formation or were introduced as contaminants during the process of obtaining the samples. Certainly some of the reports arouse grave suspicion that the organisms studied may have been contaminants. Thus Kuznetsov *et al.* comment that the data of Mekhitiyeva and Malkova, obtained by direct microscopic counts, appears to be unreliable because the extremely small number of cells actually counted might easily have arisen by contamination.[4] The same authors suggest that any samples which reveal the presence of large numbers of organisms able to grow on meat–peptone agar should be regarded as contaminated.

It is, however, most unlikely that all reports of bacteria in oil or rock samples are due to contamination. Davis and Updegraff considered the possibility that contamination was responsible for the presence of bacteria in their core samples.[1] Because many of their 162 samples were sterile they reasonably concluded that contamination was probably not responsible for the bacteria found in the others. Smirnova made a study of the penetration of drilling mud into core samples of different rock types during the drilling process (reported by Kuznetsov *et al.*[4]). A culture of *Bacterium prodigiosum* was introduced into the drilling muds and the core samples obtained were washed with water and examined in the laboratory. Here concentric layers of the rock were removed and the penetration of the bacteria assessed. In most cases bacterial penetration was limited to the outer few millimetres of the core; only when the core was fractured could deep penetration take place. Samples taken from the centre of unfractured cores 89 mm in diameter were therefore considered to give a safe indication of the bacterial content of the rock.

Kuznetsov *et al.* also cite the work of Andreyevsky who examined rock from the Ukhta oil fields where oil was extracted by mining methods.[4] Andreyevsky was able to take samples of oil, water and rock direct from the producing stratum under aseptic conditions. A large number of samples was examined: desulphurising and denitrifying bacteria were isolated in every case.

Eksertsev, cited by Kuznetsov et al.,[4] examined the bacteria in core samples from the Buguruslan and Saratov regions of the USSR. In the oil-containing strata large numbers of bacteria were detected (35–117 million/g of rock) in samples from as far down as 6560 ft (2000 m), the greatest depth examined. In most cases no bacteria were found in the over- and underlying strata suggesting that bacteria were not being introduced via the drilling mud.

A high proportion of the oil samples examined in the Russian studies also revealed bacterial activity. In one study bacterial numbers were determined in four deposits and total counts ranging from 39 000 to 558 000/ml were obtained. In another investigation samples from 16 formations were investigated by subculturing the bacteria. Most of the samples contained living bacteria, and the most common type was claimed to produce gas from oil anaerobically; *Methanobacterium omelanskii* was also identified in many of the samples.[4]

3.1.3. Bacteria in Formation Waters

There are several reports of the presence of bacteria in waters produced from oil-bearing formations. Thus Bastin and Greer and ZoBell recovered sulphate-reducing bacteria from brines, some of which were capable of growth in the presence of high salt concentrations and at temperatures as high as 80 °C.[8–10] Greve et al. found 10^4–10^5 bacteria/ml in produced waters, consisting of a mixture of aerobic and anaerobic species.[11] Gas formers and sulphate reducers were present but no sulphur-oxidising bacteria. Spurny and Dostalek also reported the presence of sulphate reducers in produced waters whereas the investigations of Lazar's group revealed a rich bacterial flora including representatives of *Bacillus, Pseudomonas, Micrococcus, Mycobacterium, Clostridium* and members of the Enterobacteriaceae.[12,13] In the deep reservoirs with high temperatures, spore-forming bacilli and cocci predominated, while in the shallow reservoirs the main types found were non-sporeformers, especially pseudomonads and members of the Enterobacteriaceae.

Kuznetsov *et al.* have provided a useful summary of some of the Russian work.[4] Several investigators report the widespread presence of bacteria in formation waters. Kuznetsov followed the changes in bacterial population along the course of an aquifer in the Dagestan deposits. From its source at the surface in the Chernyye Mountains, water passes along the permeable rock, directly under an oil deposit and emerges at the surface in the Bragunov region, giving rise to several springs. Bacteria were found at all points from which samples could be taken (at either end where the aquifer rocks outcrop at the surface, from five wells passing through the oil-bearing strata, and from two others). Bacteria were present in all samples examined in numbers ranging from 10^4 to more than 10^6/ml. Sulphate-reducing bacteria were present in the highest numbers in the vicinity of the oil; hydrogen sulphide concentration was also highest in this region.

The numbers of bacteria found in formation waters appear to be correlated with the exploitation of the deposits: Shturm (see reference 4) found that the numbers of bacteria at several sites in the Syzran and Tuymaza regions of the USSR increased substantially over eight years of exploitation of the oil deposits. Ekzertsev and Kuznetsov (see reference 4) found that methanogens, sulphate reducers and bacteria which produced gas from oil under anaerobic conditions were widely distributed in the formation waters of a number of oil deposits from the Middle-Volga region.

Reports of the isolation of bacteria from formation waters must be interpreted with some caution. It has only rarely been possible to take samples directly from the formation with appropriate bacteriological precautions and it has been suggested, for instance, that bacteria in formation waters originate by contamination with drilling muds. Where samples are taken from production wells there is a danger that the bacteria originate not from the formation itself but from the well: the casing of a well some hundreds of metres deep would provide a surface area large enough to support a considerable bacterial population. Cells becoming detached from this surface and arriving at the surface would be indistinguishable from cells originating in the formation.

Despite this note of caution it is most likely that many of the reports of bacterial activity in formation waters are reliable. Kuznetsov *et al.* point out that a rich and abundant bacterial flora has frequently been obtained from water samples immediately after drilling, before any contaminants introduced via the drilling mud could have multiplied extensively.[4] Unfortunately, to the best of our knowledge, no one has reported bacteriological data on water samples obtained immediately after drilling, together with data on the bacterial flora of the drilling mud. Again the report of Andreyevsky (see reference 4) on bacteria in formation water samples obtained aseptically during mining operations in the Ukhta oil fields clearly avoids the uncertainties mentioned above.

3.1.4.   The Origin of Bacteria in Rock, Oil and Formation Water

Given that bacteria appear to be present in rock, oil and water samples from many formations we must consider how they got there. It seems unlikely that they are a remnant of the original bacterial population dating from the formation of the oil itself. The most likely explanation for their presence is penetration along aquifers from the surface. Kuznetsov *et al.*[4] cite data of V.A. Sulin suggesting vertical rates of movement of 5m/month for water in aquifers. The penetration of bacteria from the surface would thus take many years but since such waters frequently contain appreciable quantities of organic carbon, as does the rock through which they pass, penetration accompanied by slow bacterial growth does not seem altogether impossible.

3.2.   UNDESIRABLE BEHAVIOUR OF BACTERIA IN RESERVOIRS

3.2.1.   Plugging of Formations

Present practice in the oil industry involves the injection of enormous volumes of water into reservoirs for pressure maintenance and waterflooding or as a means of disposing of produced water. On numerous occasions, microorganisms present in the water have caused plugging of the injection well, sometimes to such an extent that the well becomes useless.[1,7,14–16] Reservoir engineers have devised systems of water treatment, which are generally successful at keeping the problem under control, by methods such as filtration to remove microorganisms from the water and the addition of biocides to discourage further growth (see section 3.3.1.).

Techniques for using microorganisms in enhanced oil recovery involve the deliberate introduction of bacteria into injection wells and the encouragement of their growth in the formation. Clearly there are two dangers: that the bacteria deliberately introduced into the formation will cause plugging and that the deliberate injection of bacteria into the formation will inevitably allow some contaminating forms to gain entry as well. We do not see it as inevitable that normal bacteriological control procedures, aimed at preventing the entry of undesirable bacteria, must be suspended during the injection of the required organism. A two-stage procedure is conceivable, in which undesired species are removed from the injection waters by filtration, or killed by the addition of a short-lived biocide, and this is followed by admission of the required bacteria into the decontaminated water stream (see discussion in section 3.3.2). We shall therefore briefly consider the process of plugging by bacterial action to evaluate the dangers.

3.2.1.1. Plugging by Bacterial Cells

The ability of bacterial cells to plug porous rocks causing reduced permeability is well known. Thus Merkt[26] studied the effects on sand and limestone core samples of oilfield waters containing sulphur-oxidising bacteria, sulphate-reducing bacteria, iron bacteria and blue-green algae. Under the conditions of Merkt's experiment the core samples showed reductions in permeability ranging from 20 to 75%.

From reports in the literature it is clear that, if precautions are not taken, very considerable quantities of bacterial and algal cells together with their extracellular slimes and precipitates are likely to be introduced into the formation during waterflooding.[7,14,16-21] Bacteria can be removed from water by treatments such as flocculation and filtering; if required it is possible to achieve virtually complete removal of bacterial cells but economic considerations dictate the employment of treatments sufficient to reduce bacterial populations to acceptable levels coupled with the use of biocides to prevent growth. Unfortunately, it is not a simple matter to decide, in general, what constitutes an acceptable level of bacterial cells. Sharpley claimed that rocks with permeability values in the darcy range would not plug with an injection of water containing 10^6 cells/ml, while with rocks in the 20 mD range the cell concentration should be less than 3×10^4 cells/ml.[22] Rocks with permeabilities in the range 1–2 mD required water virtually free from bacteria in order to avoid plugging. Although this figure was certainly applicable to the situations from which it was derived, it is doubtful if it has general validity. On the contrary, Allread said that no general law could be laid down relating the bacterial concentration to its effect on permeability because every reservoir is different in important respects.[16]

Plugging of the formation by the cells actually introduced is not the only danger: growth may take place either on the well injection face or in the formation itself, and plugging may be caused either by the larger cell population or by the products of its metabolism. This is illustrated by data cited by Allread on the injection of sea water in secondary recovery operations in California.[16] The water contained a high bacterial population (2–15×10^7 cells/ml) and reduced the permeability of core samples by 80%. However, in the presence of mercuric chloride no reduction in permeability occurred, showing that the effect was not due to the bacterial cells originally present in the sea water but depended either on growth or perhaps the products of metabolism. This emphasises the need not only to reduce the bacterial population in the flood water to a low level but to prevent subsequent growth and metabolism, a function fulfilled by the addition of biocides to the flood water in commercial practice.

3.2.1.2. Plugging by Products of Bacterial Metabolism
Various types of bacteria can produce metabolic products likely to cause formation plugging. Plummer *et al.* mentions ferric hydroxide, metallic sulphides, sulphur, calcium carbonate, and gelatinous and chitinous materials as bacterial products likely to cause blocking.[21] The presence of air in bodies of water above ground permits a large variety of organisms to grow and produce metabolic products and, although these may be removed by precipitation or filtration, there is still a danger of growth in pipework systems and in the well itself. Slime-forming members of the genus *Pseudomonas* have caused serious plugging problems in some American fields apparently growing best at points of stagnation in pipework.[16]

Another group of organisms, the iron bacteria,[23] are capable of causing serious trouble.[16,22] In waters containing iron and oxygen they grow and produce voluminous precipitates of ferric hydroxide. Under the right conditions huge gelatinous masses of ferric hydroxide can form on pipe walls and can clog the pipes or slough off and plug the well-face. Slime masses of iron bacteria, 45 cm thick, have been observed in water injection systems.[16] Iron bacteria do not appear to be salt tolerant: no growth has been observed when brines have been used as flood waters.

The organisms most often responsible for plugging are the sulphate-reducing bacteria, members of the genus *Desulfovibrio* and the genus *Desulfotomaculum*. These organisms also seem to be more frequently detected than any others by microbiologists investigating rock, oil and formation water samples: it is not clear whether this reflects their natural abundance or the interest shown in them by microbiologists aware of their undesirable properties. The sulphate-reducing bacteria are obligate anaerobes: they can grow only in the virtual absence of oxygen but they survive in a wide variety of apparently oxygenated environments, and having no requirement for oxygen, are capable of growth in anaerobic situations in wells and in formation rock. Their normal metabolism is based on the reduction of sulphate, which serves as an electron acceptor, to sulphide. They can obtain much of their carbon by reduction of carbon dioxide although a source of organic carbon is essential. The range of organic carbon compounds known to be utilised is very restricted: in laboratory culture they grow most rapidly on carboxylic acids such as lactate, pyruvate, malate and fumarate, although slower growth is possible on a few sugars and alcohols. Hydrogen can be oxidised by the enzyme hydrogenase. A useful review of the physiology of sulphate-reducing bacteria has been published.[24]

Sulphate is commonly present in formation waters and it is clear that sulphate-reducing bacteria can and do grow in reservoirs. Formation waters are unlikely to contain appreciable quantities of the carbon sources utilised in laboratory culture; possibly these are supplied by other microorganisms growing in the reservoirs. During growth sulphate is reduced to sulphide; this combines with any ferrous ions present to form a precipitate of ferrous sulphide which together with the ferrous hydroxide also formed can have an adverse effect on permeability.

The problem of plugging is discussed further in section 3.4.

3.2.2. Degradation of Oil in Reservoirs

In view of the evidence that bacteria are present in at least some oil-bearing formations some microbiologists have considered the possibilities that they might degrade some components of crude oil *in situ*. The Russian work on this topic has been summarised by Kuznetsov *et al.*[4] They point out that many analyses have shown that waters in direct contact with oil formations contain large quantities of methane and its homologues. The methane content decreases with increasing distance from the oil, suggesting a constant outward diffusion of methane. Some authors have considered that this implies a constant replacement of methane by breakdown of some oil component. Ginzburg-Karagicheva and Ekzertsev (see reference 4) have attempted to demonstrate such breakdown in the laboratory under anaerobic conditions by incubating crude oil and core samples from various formations in stoppered tubes with ammonium phosphate solution as a source of nitrogen and phosphorus. These tubes were, in turn, enclosed in outer tubes filled with liquid and also stoppered, presumably to exclude air from the inner tube. After periods of two to four weeks gas bubbles of up to several millilitres in volume were formed in the inner tubes. About 20 strains of bacteria were isolated from tubes in which gas had formed. Two were described as closely resembling *Methanococcus mazei* and *Sarcina methanica*. The 20 isolates were themselves capable of generating gas under anaerobic conditions in the presence of sterilised crude oil and ammonium phosphate solution. The composition of the gas formed was broadly similar in all the experiments and contained, by volume, 56–78% nitrogen, 15–40% methane and 2–5% carbon dioxide. Hydrogen was absent in most samples but comprised up to 5% of the volume in a few cases.[4]

Observations of this sort strongly suggest that decomposition of oil *in situ* is a real possibility and it is both unfortunate and surprising that no detailed metabolic studies appear to have been described. As with other reports of anaerobic hydrocarbon degradation, assessment of their significance must await more detailed studies.

The formation of nitrogen as the principal product in the experiments described above is of interest because nitrogen can form a considerable proportion of natural gas samples.[4] Andreyevsky has suggested that the anaerobic decomposition of oil might occur via a mechanism involving nitrogen bridges and that this might account for the high nitrogen content of some petroleum gases. Natural gases also contain argon and this fact has suggested to some authors that the nitrogen present must have been derived from air, which contains about 1% argon, rather than by bacterial decomposition of oil.[27,28] However, measurements of the argon : nitrogen ratios in natural gas samples revealed values which, in most cases, were considerably lower than that in air indicating that much of the nitrogen present was derived from oil breakdown. In any case the argon present could have been of geological origin.

If microorganisms were indeed active in oil-bearing formations the consequences might be serious. One author has suggested that such activity is widespread and that some 10% of the world's oil resources have been destroyed by bacterial action.[29]

We must also consider whether there is a danger that a microbial culture, introduced into a formation with the intention of enhancing oil release, might actually decrease recovery by bringing about unfavourable changes in the oil. The most likely possibility would be that of an increase in viscosity making oil expulsion by the waterflood more difficult. A few reports do exist of changes in oil properties during field trials of microbial enhancement techniques but there is nothing to suggest that a deleterious effect on oil properties is to be expected. Both Jaranyi and Karaskiewicz have reported marked decreases in the viscosity of oil in their field trials;[30,31] and although Kuznetsov et al. found that the viscosity of oil increased, their trial resulted in at least some increase in oil production so the effect was presumably not serious.[4] Clearly in any future trials, especially in the case of those involving fields still considered to have some commercial value, any effects on oil viscosity would be determined in advance by laboratory measurements. The use of an organism with the ability to attack only specific components of the crude oil would make such prediction much easier.

3.2.3. Bacterial Corrosion

The promotion of metallic corrosion by microorganisms is a well-known phenomenon which was first recognised by Gaines in 1910.[32] Some idea of the magnitude of the problem can be gained from the estimates that in the United Kingdom at least 50% of failures of underground pipes are due to microbial action[33] and that in 1955 the maintenance and replacement of underground pipes because of corrosion cost about £20 million.[34]

The oil industry suffers considerable losses due to the anaerobic corrosion of iron and steel structures and many of these losses are due to the activities of bacteria. Because of the widespread distribution of sulphate-reducing bacteria and sulphate in natural waters, iron and steel in closed systems are commonly subject to anaerobic corrosion.[35,36] Doig and Wachter have documented an example of anaerobic bacterial corrosion in California.[37] In a series of failures of oil-well casings, localised perforations occurred in the $\frac{3}{8}$ in (9 mm) steel pipe at depths between 900 and 7000 ft (270–2100 m). The pipe corroded through in different positions in times averaging about four years. Sections of the casing showed iron sulphide accumulation. Large numbers of sulphate-reducing bacteria were present in the drilling mud.

Allread refers to serious corrosion in producing wells in Indiana accompanied by high counts of sulphate-reducing bacteria.[16] Over a two-year period a chlorinated phenol derivative was introduced to suppress bacterial growth. Failures in pumps, rods and tubing were reduced by 40–80% and corrosion rates, as measured by corrosion coupons, were also reduced considerably. The chemical control agent itself had no corrosion control properties so this clearly represented a case of bacterial corrosion.

A number of different mechanisms of bacterial corrosion have been recognised; these include cathodic depolarisation, the production of corrosive metabolites, the disruption of natural or artificial protective films, the formation of different aeration or concentration cells by the growth of bacterial colonies on the metal surface and the breakdown of corrosion inhibitors. In each case the fundamental action is an electrochemical one, the role of the bacteria being to create conditions favouring the electrochemical process. A review of microbial corrosion has been published.[38]

Bacteria can promote corrosion under both aerobic and anaerobic conditions and a variety of species may be involved. In the oil industry, anaerobic conditions predominate in wells and oil-bearing formations and the bacteria most frequently involved in corrosion are the sulphate reducers. The discussion here will deal mainly with this group.

The first suggestion as to the mechanism of corrosion of iron pipes by sulphate-reducing bacteria was made by von Wolzogen Kühr and Van der Vlugt.[39] They suggested that corrosion was promoted because the bacteria could bring about cathodic depolarisation by oxidising the protective layer of hydrogen using the enzyme hydrogenase. The scheme of reactions proposed was as follows:

1. $4Fe \longrightarrow 4Fe^{2+} + 8e^-$ (At the anode)
2. $8H_2O \longrightarrow 8H^+ + 8OH^-$
3. $8H^+ + 8e^- \longrightarrow 8H$ (At the cathode)

4. $SO_4^{2-} + 8H \longrightarrow S^{2-} + 4H_2O$ (Cathodic depolarisation by bacteria)

5. $Fe^{2+} + S^{2-} \longrightarrow FeS$

6. $3Fe^{2+} + 6OH^- \longrightarrow 3Fe(OH)_2$

Sum: $4Fe + SO_4^{2-} + 4H_2O \longrightarrow 3Fe(OH)_2 + FeS + 2OH^-$

There is considerable evidence that cathodic depolarisation by hydrogen oxidation is in fact involved in corrosion by sulphate-reducing bacteria. For example Horvath and Solti[40] demonstrated cathodic depolarisation of mild steel electrodes in cultures of sulphate-reducing bacteria and Booth and Wormwell showed that the rate of mild steel corrosion by a series of cultures of sulphate reducers was proportional to their respective hydrogenase activities.[41] However, the mechanism suggested above would predict a molar ratio between iron corroded and ferrous sulphide produced of 4:1. In practice, ratios varying between 1:1 and 48:1 have been reported, suggesting that the actual mechanism is more complex than that suggested above, and it is clear that mechanisms other than cathodic depolarisation by hydrogen oxidation are involved. For instance, the ferrous sulphide produced may itself play a part in the corrosion. This was demonstrated by Booth et al. who showed that chemically-prepared ferrous sulphide itself caused cathodic depolarisation in the absence of bacteria.[42] There may well be other mechanisms involved in corrosion by sulphate reducers but the two mentioned are probably the most important.

3.2.4. Souring

The production of sulphide in oil-bearing formations by sulphate-reducing bacteria can have another undesirable consequence: the oil and gas produced can become soured, i.e. contaminated with hydrogen sulphide which may cause corrosion and require expensive treatment to remove it prior to sale. In one instance of which we are aware sea water containing sulphate-reducing bacteria was used for flooding a Californian well. After a short while the gas produced became so contaminated with hydrogen sulphide that it was unsaleable. Recovery on changing back to a fresh-water flood took several years.

3.3. MEASURES TO CONTROL BACTERIAL GROWTH IN RELATION TO ENHANCED OIL RECOVERY TECHNIQUES

3.3.1. Types of Control Measure

The various undesirable activities of bacteria in reservoirs frequently necessitate treatment of the injection water to control growth in connection with secondary or tertiary recovery operations. Injection water will also in many cases be subjected to other treatments. A scheme of water treatment might include the following stages:

1. Treatment with a chemical flocculating agent to remove solids;
2. Deaeration and addition of biocide (to minimise bacterial growth on the filters);
3. Passage through a sand or diatomaceous earth filter;
4. Passage through a cartridge filter en route to a holding tank;
5. Passage through a polishing filter immediately prior to injection;
6. Additions of substances such as biocides, corrosion inhibitors, oxygen scavengers;
7. Injection.

Numerous proprietary products are available as additives and some care is necessary in selecting combinations of substances which are compatible and which do not have undesirable side effects. Thus acrolein (added as a biocide) will react with sulphite (added as an oxygen scavenger) with the result that neither function is served. Phosphate is sometimes added as an anodic inhibitor of corrosion; since in some cases bacterial growth in formations is limited by shortage of phosphate this addition may serve to promote bacterial growth.

A variety of substances have been used as biocides. Chlorine, long used in the purification of water for domestic purposes, has been employed in many instances. Beck has pointed out the limited usefulness of chlorine in some applications.[17] He reported that although bacterial counts could be reduced to very low levels, injection rates usually decreased over a period of time. In one test on a well in which chlorination had been practised for two years water was allowed to backflow for 3 h at which time over 20 000 living bacteria/ml were present. Chlorine is ineffective because of its high reactivity; when organic material is present its concentration rapidly decreases.

Formaldehyde is an effective inhibitor of sulphate-reducing bacteria in laboratory cultures.[17,43] It has also been used in field operations with some success.[4,36,44] Many other types of compound have found favour, including chromate ion (which also has anticorrosive properties), acrolein, quaternary ammonium compounds, amines and chlorinated phenols. No single compound is useful in all situations. All have their advantages and limitations. Quaternary ammonium compounds, for instance, are very effective in fresh water but their activity is greatly reduced in brines. Some organisms, especially pseudomonads, can rapidly become resistant to their action and can then cause serious plugging. Amines can also be very effective but due to their strong absorption characteristics and surface activity can themselves cause plugging.[16] The chlorinated phenols are effective against sulphate-reducing bacteria but are much less soluble at low pH. Their use is restricted by law in many countries because of their general biological toxicity. Chromate and other metallic ions may contaminate the oil and cause undesirable catalytic effects.

Tertiary recovery operations frequently involve the injection of expensive biodegradable components such as polysaccharide viscosifiers. In these circumstances the activity of unwanted bacteria, which can degrade the polysaccharide and use the products as a carbon source for growth, poses particularly severe problems. A biocide for use in such situations needs, in addition to the normal properties, the ability to inhibit bacterial action in the region of the polymer for extended periods of time. It must be chemically and biologically stable, must not be adsorbed on to the rock or partitioned into the oil and must co-migrate with the polymer. At present no ideal candidate for this task seems to exist.

Corrosion is frequently caused by the activities of microorganisms in drilling muds. In cases where biodegradable components such as starch or carboxymethylcellulose are present, bacterial activity can reduce the efficiency of the mud. Apart from the addition of biocides, control may be achieved by adjustment of pH to a value above about 9.[45]

3.3.2. The Growth of Undesirable Bacteria During Enhanced Recovery Operations Using Microorganisms

Since the growth of unwanted microorganisms in oil fields can have serious consequences and it is common practice to take precautions against both the introduction of bacteria with the waterflood and their subsequent growth, we must consider how far these precautions would be compatible with enhanced recovery operations using microorganisms. Clearly the bacterial inoculum would be introduced into the waterflood at the latest possible stage to avoid removal by any system of filtration and the main problem would be caused by the effects of the various substances (biocides, oxygen scavengers, corrosion inhibitors) added to the flood water.

One strategy would be to add the bacterial inoculum in a discontinuous manner, suspending the addition of inimical substances for a period before, during and after the addition of the bacteria. Most, at least, of the field trials reported so far have used discontinuous inoculation and it is possible that such a method would be acceptable in operations on a much larger scale. There would, however, appear to be a serious danger that the biocide and the bacterial inoculum, although added separately, would migrate through the formation at different rates and that the bacteria would be killed or their growth inhibited. We therefore do not think this is likely to be a workable strategy. As an alternative to adding biocide discontinuously it might be possible to achieve an acceptable reduction of the bacterial count by treating the flood water with a biocide such as chlorine and inactivating it prior to the addition of the bacterial inoculum. In some circumstances it might be possible to dispense with the addition of biocides: none of the field studies described so far has mentioned their use. However, the need for biocides is likely to be dictated by factors such as the level of bacterial contaminants in the flood water and its nutrient status; in many cases no choice will be possible if the field is to be properly exploited.

Perhaps the most promising approach would be to use a bacterial culture which was resistant to the biocides and other additives used. Methods of obtaining resistant cultures are considered in Chapter 5. The chance of success would depend on a careful choice of the additives. For instance, with some biocides such as chlorine, resistant strains are not known; on the other hand resistance to quaternary amines is comparatively easily acquired.

A problem might arise due to plugging by the cells used as an inoculum. This would be minimised if growth occurred only when the cells had penetrated well into the reservoir. The strategy of using oil or some component as the carbon source for growth would dispense with the need for a complete spectrum of nutrients in the waterflood and should achieve this result automatically. Since the injection face and the formation rock immediately adjacent to it is undoubtedly the region most vulnerable to plugging, elimination of growth in this region should prevent plugging by the inoculum.

3.4. PENETRATION OF BACTERIA THROUGH ROCK FORMATIONS

3.4.1. The Problem of Penetration

A critical factor in the use of microorganisms for enhanced oil recovery is their ability to penetrate extensively in the formation rock. With some of the possible enhancement mechanisms we visualise, e.g. production of surfactants *in situ* by bacterial metabolism of crude oil, it might be sufficient for bacteria to penetrate only part of the way into the swept area of the field, the surfactants produced being carried to the rest of the field by the waterflood. Other possible mechanisms, such as the ejection of oil from blind capillaries by *in situ* generation of gas, would require a very intimate and complete penetration by the bacteria. It would not be necessary for success that all the bacteria should move freely through the formation since in any case we envisage providing an inoculum which would grow at the expense of the oil-in-place; it might be sufficient if a single cell were to penetrate to a particular site. If oil release were dependent upon the combined activities of two or more strains of bacteria, the problem would be more critical since the required action would depend on the simultaneous presence of all the strains at the same site.

Those bacteria which fail to penetrate through the rock may cause problems by obstructing the water flow along fine channels and interfering with the waterflood (see section 3.2.1.). The production of metabolites such as iron sulphide or ferric hydroxide would also be a potential cause of plugging (but see section 5.2.2.). There would thus be a potential danger to the functioning of the reservoir from injected bacteria and any programme using microorganisms to enhance oil recovery would need to assess the dangers and devise appropriate control measures. Unfortunately little information is available about the ability of bacteria to penetrate reservoir rock but it is clear that several variables are likely to be important. These include characteristics of the bacteria (size, shape, deformability, the nature of the cell surface, whether or not cells aggregate into clumps or chains, the formation of extracellular substances such as sheaths and slimes, and the ability to swim by the use of flagella); reservoir characteristics such as permeability, pore structure, the chemical nature of the rock (which may determine the tendency of cells to adsorb on to the surface), chemical composition of the connate water and of the flood water (which may affect clumping and may also influence adsorption on to surfaces) and temperature and pressure (which may restrict bacterial swimming movements and may also change bacterial morphology). The rate of water movement through the porous rock, the pressure gradient and the concentration of bacteria in the water might also be important.

The situation is thus potentially very complex although many of the above variables may yet turn out to have comparatively little effect on penetration. Ideally three types of information are needed as a basis for work on enhanced oil recovery: we need to know about the relative importance of the above variables and, in particular, the range of values within which operations might be feasible; we need to be able to estimate bacterial movement rates through porous rock under optimum or near optimum conditions in order to calculate the minimum time needed to penetrate with specific well spacings, and ideally we need a model which can predict the movement of bacteria through the reservoir under specified conditions.

3.4.2. Laboratory Measurements of Penetration

These problems are susceptible to study under controlled laboratory conditions (although the number of experiments required to build up a picture involving all the variables mentioned above is somewhat daunting), but very little published information is available. So far most studies have employed cylindrical cores of outcrop rock or sandpacks to measure penetration by the test organism. These are usually presaturated to the desired degree with oil and water. One serious problem is that of obtaining a series of identical cores: even cores cut from the same piece of rock are likely to vary considerably in their permeability with the result that valid comparisons are difficult to make. Even when sandpacks are made up from the same material and to the same permeability, the rates of migration of bacterial spores through them can differ considerably (Yarbrough, personal communication).

Several different methods of measuring rates of penetration through the core or sandpack have been used. Yarbrough (personal communication) performed experiments with sandpacks fused into lucite tubes. The test organism was *Clostridium acetobutylicum* and spores were used to avoid the complication of increasing numbers during the experiment due to growth. A series of ports permitted samples to be taken so that estimates could be made of the distribution of spores along the length of the sandpacks at different times. Fekete studied both penetration and plugging using dead and living cells of *Bacillus subtilis* in cores of Berea sandstone 7–9 cm in length.[25] Penetration and plugging were assessed mainly by measuring water flow and the pressure drop across sections of the core. The pressure applied across the short core (40 psig) was rather high compared with the pressure gradients likely to occur in reservoirs, and the choice of bacterial strain was somewhat unfortunate: the cells were large (ca. $4 \times 1 \mu m$) and tended to form chains and produce slime, thus making plugging much more likely.

The experiments of Myers and McCready are of particular interest.[46] They worked mainly with cores of Berea sandstone up to 36 cm in length but also performed some experiments with other rock types. Their experimental organism was *Serratia marcescens* which was labelled prior to use by growth in the presence of ^{32}P-labelled phosphate. The progress of bacterial penetration could be monitored during the experiment by measuring the emission by radioactivity at different points along the core. Afterwards the core was split in two longitudinally, one half was examined by autoradiography, the other half was examined by culturing *Serratia* from it. Both methods gave essentially the same results although it was notable that penetration was rather uneven.

3.4.3. Data on Bacterial Penetration

3.4.3.1. Rates of Penetration
Several estimates of absolute rates of movement have been published or can be calculated. Beck and ZoBell both estimated the rates of penetration of sulphate-reducing bacteria through sandpacks to be of the order of 2·5 cm/day, a rate which would appear to be too slow to be useful.[10,47] From the data of Fekete it is clear that his strain of *Bacillus subtilis* could penetrate through Berea sandstone at about 10 cm/day and Myers and McCready found very similar rates of movement of their test organism, *Serratia marcescens*, through the same type of rock.[25,46]

It is also useful to consider the ratio between the linear rate of water movement and that of bacteria. Myers and McCready found a ratio of about 1·8 with *S. marcescens* moving through a core of Berea sandstone and Yarbrough (personal communication) found ratios ranging from 1·3 upwards when measuring the movement of *Clostridium acetobutylicum* through sandpacks.[46]

What do these figures mean in terms of field operations? If the higher of the estimates for rates of movement were to apply it would take about 2·5 years for bacteria to move between an injection well and a production well 90 m apart. If the spacing were 450 m the time taken would be about 12·5 years. If we assume a mean linear rate of water movement of 30 cm/day, and assume that water moves 1·8 times as fast as the bacterial cells,[46] the times taken for bacteria to penetrate 90 m and 450 m would be about 1·5 years and 7·5 years, respectively. These estimated times are long but it must be remembered that they do not represent the time delays before the first effects could be detected in field operations. These would be rather shorter and would presumably be the respective times taken for bacteria to penetrate into the first part of the field plus the times taken in each case for mobilised oil to flow to the production well. The times calculated above might represent the delays before any effects of bacterial action could be developed throughout the entire swept volume under consideration. Viewed in this light they do not seem entirely discouraging.

We must also take into account the limited data available from actual field trials. These are quite encouraging. Thus Jaranyi mentions the recovery of bacteria from a production well 1·6 km from his injection well only four months after injection, although in this case the limestone reservoir rock had been fractured.[48]

In the field trial conducted by Coty, Updegraff and Yarbrough (Yarbrough, personal communication) the injection water took 70 days to arrive at the production well 120 m away, products of bacterial metabolism began to arrive after 80 days and oil production increased significantly after 4 months. No bacterial breakthrough occurred during the experiment which lasted almost 12 months. This confirms that beneficial effects can occur well before the bacteria have completely penetrated the formation, suggesting that in a successful large-scale operation increased oil production could be expected in considerably shorter times than those calculated above.

Although the circumstances of field trials are more realistic than those of laboratory experiments there are still many difficulties in interpreting figures for rates of bacterial movement. Thus bacteria emerging at a production well might get there by rapid passage along a rock fissure and their rate of movement might be quite uncharacteristic of that through the mass of the rock. This would be especially likely where fracturing had occurred as in the case cited by Jaranyi.[48] Perhaps the most useful information on rates of bacterial penetration to be obtained from field trials is that penetration, in many cases at least, has been sufficiently rapid to cause measurable effects on oil production.

3.4.3.2. The Effects of Rock Permeability

A number of authors have reported data on plugging or penetration of bacteria with sandpacks or core samples of different permeabilities and some comments have been based on field experience. Jaranyi (personal communication) suggested that microorganisms could only be used in formations with permeabilities in excess of 60 mD because of the dangers of plugging while in a publication Jaranyi et al. categorised formations into low (10–80 mD), medium (150–300 mD) and high (400–700 mD) permeabilities, suggesting that only the latter two groups would be accessible to microbial techniques.[49] Rader (personal communication) found that *Escherichia coli* failed to penetrate a core with a permeability of 80 mD. Yarbrough (personal communication) observed penetration of *Clostridium acetobutylicum* through sandpacks with permeabilities down to 260 mD but not at 170 mD. Davis and Updegraff reported their results in slightly different terms.[1] They found that sulphate-reducing bacteria up to about $0.6 \times 3 \mu m$ in size would pass through rocks which contained an appreciable number of pores larger than $3 \mu m$ without causing serious plugging. On this basis they concluded that small bacteria should be able to penetrate many reservoir rocks without plugging problems.

Fekete, although primarily interested in plugging rather than penetration, found that *Bacillus subtilis* penetrated the entire length of all of his cores of Berea sandstone.[25] He could not detect any consistent effect of permeability on the extent of plugging (range 64–301 mD). Myers and McCready found penetration through all their rock cores.[46] These included Berea sandstone

(100–300 mD), Mississippian limestone (<0·1 mD), Early Devonian limestone (1·7 mD) and Late Mesozoic sandstone (<0·1 mD) suggesting that low permeabilities need not necessarily rule out the use of microorganisms.

The reports from different authors are thus somewhat conflicting although it must be remembered that this may be partly due to some authors considering penetration and some considering plugging. Although these are related they are not the same thing. It is perfectly possible in principle for bacteria to penetrate through a formation quickly but, by growth or metabolism, to reduce the permeability to an unacceptable degree.

Finally, we must point out that, at present, we are far from being able to construct a mathematical model to predict penetration and plugging by bacteria in a reservoir. Of the many variables which might be involved, only a few have been studied at all, and there is no general picture of the significance even of these.

REFERENCES

1. Davis, J. B. and Updegraff, D. M. (1954). *Bacteriol Rev.*, **18**, 215.
2. Knösel, D. and Schwarz, W. (1954). *Arch. Mikrobiol.*, **20**, 362.
3. Wang, H. S. and Schwarz, W. (1961). *Z. Allgemein. Mikrobiol.*, **1**, 223.
4. Kuznetsov, S. I., Ivanov, M. V. and Lyalikova, N. N. (1963). *Introduction to Geological Microbiology.* (English translation ed. C. H. Oppenheimer) McGraw Hill, New York and London.
5. Kuznetsov, S. I. (1965) Abh. Deut. Akad. Wiss. Berlin, Kl. *Chem. Geol. Biol.*, 173 (published 1966).
6. Schwartz, W. (1972). *Naturwissenschaften*, **59**, 358.
7. Beerstecher, E. (1954). *Petroleum Microbiology.* Elsevier, New York.
8. Bastin, E. S. (1926). *Science*, **63**, 21.
9. Bastin, E. S. and Greer, F. E. (1930). *Bull. Am. Assoc. Petrol. Geologists*, **14**, 153.
10. ZoBell, C. E. (1947). *Oil and Gas Journal*, **46**, 62.
11. Greve, W., Müller, A. and Schwarz, W. (1957). *Zentr. Bakteriol. Parasitenk, Abt.* II. **110**, 82.
12. Spurny, M. and Dostalek, M. (1956). *Pracé ustavu pro naftory vyzkum* Publ. No. 26–30, 59.

13. Lazar, I. (1979). In: *European Symposium on Enhanced Oil Recovery, 1978* ed. J. Brown, p. 279. Institute of Offshore Engineering, Heriot-Watt University, Edinburgh.
14. Davis, J. B. (1967). *Petroleum Microbiology*. Elsevier, New York.
15. Allison, L. C. (1947). *Soil Science*, **63**, 439.
16. Allread, R. C. (1976). In: *The Role of Microorganisms in the Recovery of Oil. Proceedings of the 1976 Engineering Conference*, p. 133. National Science Foundation, Washington.
17. Beck, J. V. (1947). *Producers' Monthly*, **11**, 21.
18. Cerini, W. F., Battles, W. R. and Jones, P. H. (1946). *Petroleum Technol.*, **9**, 2028.
19. Lada, A. (1959). *Oil and Gas Journal*, **57**, 93.
20. Raleigh, J. T. and Flock, D. L. (1965). *J. Petroleum Technol.*, **17**, 201.
21. Plummer, F. B., Merkt, E. E., Power, H. H., Sawin, H. J. and Tapp, P. F. (1944). Am. Inst. Mining Met. Engrs. Tech. Pub. No. 1678.
22. Sharpley, J. M. (1961). *Petroleum Engr.*, **33**, 55.
23. Mulder, E. G. (1964). *J. Appl. Bacteriol.*, **27**, 151.
24. Legall, J. and Postgate, J. R. (1973). *Adv. Microbiol. Physiol.*, **10**, 81.
25. Fekete, T. (1959). *M.Sc. Thesis*, University of Edmonton, Alberta.
26. Merkt, E. E. (1943). *M.Sc. Thesis*, University of Texas.
27. Belousov, V. V. (1937). *Ocherk Geokhimii Prirodnykh Gazov*, ONTI, (via Kuznetsov *et al.*, ref. 4).
28. Kozlov, A. L. (1950). *Promlemy Geokhimii Prirodnykh Gazov*, Moscow, Gostoptekhizdat (via Kuznetsov *et al.*, ref. 4.).
29. Anon. (1972). *World Oil*, 28.
30. Jaranyi, I. cited by Lazar, I. (reference 13).
31. Karaskiewicz, J. (1968). *Nafta (Krakow)*, **24**, 198.
32. Gaines, R. H. (1910). *Ind. Eng. Chem.*, **2**, 128.
33. Booth, G. H. (1964). *J. Appl. Bacteriol.*, **27**, 174.
34. Vernon, W. H. J. (1957). In: *The Conservation of Natural Resources*, p. 105. Institution of Civil Engineers, London.
35. Beck, J. V. (1946). *Producers' Monthly*, **10**, 29.
36. Latter, F. (1949). *Producers' Monthly*, **13**, 47.
37. Doig, K. and Wachter, A. P. (1951). Annual Meeting, National Ass. Corrosion Engineers. Reported by H. D. Ralph (1951). *Oil and Gas Journal*, **50**, 69.
38. Miller, J. D. A. (1971). *Microbial Aspects of Metallurgy*. Medical and Technical Publishing Co. Aylesbury.
39. Wolzogen Kühr, C. A. H. Von and Van Der Vlugt, L. S. (1934). *Water, Den. Haag*, **18**, 147.

40. Horvath, J. and Solti, M. (1959). *Werkstoffe und Korrosion*, **10**, 624.
41. Booth, G. H. and Wormwell, F. (1962). In: First International Congress on Metallic Corrosion, London, p. 341. Butterworth, London.
42. Booth, G. H., Elford, L. and Wakerley, D. S. (1968). *Brit. Corrosion J.*, **3**, 242.
43. Plummer, F. B. and Walling, I. W. (1946). *Trans. Am. Inst. Mining Met. Engrs. Petroleum Division*, **165**, 64.
44. Menaul, P. L. and Dunn, T. H. (1946). *Trans. Am. Inst. Mining Met. Engrs. Petroleum Division*, **165**, 26.
45. Hunter, J. B., McConomy, H. F. and Weston, R. F. (1948). *Oil and Gas Journal*, **47**, 249.
46. Myers, G. E. and McCready, R. G. L. (1966). *Canad. J. Microbiol.*, **12**, 477.
47. Beck, J. V. (1947). *Producers' Monthly*, **11**, 13.
48. Jaranyi, I. (1968). *M. All. Földtani Intezet Evi Jelentese Az 1968, Evröl*, 423.
49. Jaranyi, I., Kiss, L. and Szalanczy, G. (1965). Abh. Deut. Akad. Wiss. Berlin, Kl. *Chem. Geol. Biol.*, 69 (Published 1966).

CHAPTER 4

Experience of Bacterial Enhanced Recovery Techniques

There is a continuing interest in the bacteriology of oil production on the part of the operators because of the possibility of contamination with undesirable forms. These may cause plugging of the formation, or of a waterflood injection face, either by accumulation of bacterial cells in small orifices or by the production of slimes or inorganic precipitates. Furthermore, deleterious effects such as souring of the crude may occur either in the reservoir or in pipelines or tankers if the oil contains appropriate quantities of water and is contaminated by certain bacteria. We are not in this discussion primarily concerned with problems resulting from such contamination, but since we are concerned with the deliberate use of bacteria for enhanced recovery it will be necessary also to consider the risk and nature of unwanted side effects.

4.1. A BRIEF HISTORY

With the realisation that bacteria could survive and grow under some reservoir conditions, and with the development of a number of schemes for chemical flooding procedures, it came to be realised that since bacteria may themselves act as 'chemical factories' it might be effective to achieve a chemical flood by using bacteria to generate the chemicals *in situ*. Perhaps the first suggestion along these lines was made by Beckman in 1926, but for nearly 20 years after that no serious thought seems to have been given to the matter.[1] Since the end of the Second World War there have been two successive phases of activity. The first was in the United States; this was moderately intensive for 10 years, less so for the next decade, but little or no work has been carried on there since the mid 1960s until the current resurgence of activity reported in Chapter 7. The impetus for these studies may have been partly derived from the burgeoning of microbiology at that time, the exhaustion of a number of fields known still to contain oil-in-place and an increase in general in awareness of enhancement of recovery. Some development also took place in the USSR in the early 1950s, but this seems to have faded away. The second phase was based upon the earlier American studies and took place in Eastern Europe, successively in Czechoslovakia (mid-1950s), Hungary and Poland (more or less coincidentally right through the 1960s) and finally in Romania in the past few years. Activity in each of these countries except Romania has now apparently ceased, in spite of some rather promising results, albeit on a small scale, from Poland. We shall discuss these in due course. The Eastern European interest may have been stimulated by a need to maximise the yields from rather limited local resources, coupled with a knowledge of the earlier American work and the progressive exhaustion of wells in use.

4.2. THE ORIGINAL IDEAS

The pioneer studies were carried out between about 1943 and 1953 by ZoBell.[2-4] Initially attention was concentrated on a number of *Desulfovibrio* strains. It was suggested that acids and carbon dioxide produced by sulphate-reducing bacteria could lead to increased rock porosity, while the excretion of detergent substances and the conversion of some higher molecular weight hydrocarbons in the crude to substances of lower molecular weight would reduce its viscosity. The culture media proposed were simple ones containing, for example, salts, calcium carbonate and calcium lactate, or some other organic acid. It was found that the acids produced might include acetic, propionic and butyric, as well as carbonic: these would react with calcium and magnesium carbonates in the rock to produce some degree of dissolution and facilitate oil flow. Evidence was obtained for the production of surfactants which may be preferentially adsorbed into sand particles and rock, and by there reducing the oil/water surface tension tend to displace oil from rock adhesion. Among the gases produced under appropriate conditions, carbon dioxide and methane would increase the local pressure in the formation and also dissolve in the oil, increase its volume and decrease its viscosity. Bacteria furthermore showed a tendency to adhere to solid surfaces and by undergrowing the oil film tend to push it off. In later work ZoBell introduced the idea of using heterotrophic bacteria able to ferment carbohydrates anaerobically with the production of gas and other useful products.[4] Underlining the value of some of these observations for oil recovery, it was observed that several strains of bacteria would grow at temperatures up to nearly 88 °C, and would migrate through unglazed porcelain, 1·25 cm thick, in a few hours, and through tightly packed sand at a rate of more than 2·5 cm/day.

4.3. THE FIRST DEVELOPMENT OF ZOBELL'S WORK

ZoBell's work generated development along these general lines by Updegraff and co-workers, and by Beck in the United States; bacteria other than sulphate reducers were examined, including some species of *Clostridium* and *Pseudomonas*.[5-8] By no means all of these studies were encouraging. Beck, for example, found in his laboratory experiments that sulphate-reducing bacteria used crude oil extremely slowly, and while some growth took place in reservoir water, no oil was released in model systems.[8] The penetration of organisms through sand cores at 2·3 cm/day was judged too slow to be of potential value, and no field trials were undertaken. Interest began to quicken outside the US. In the Soviet Union Kuznetsov found that bacteria in certain oil–gas strata in the Saratov and Buguruslan areas were present in sufficient numbers to liberate 2 g of carbon dioxide/day/tonne of rock.[9] He suggested that methane was probably formed at rock surfaces by the interaction of carbon dioxide and hydrogen. Other Soviet work has come from Andreyevsky and Shturm, while Senyukov *et al.* used aerobic and anaerobic bacteria for the tertiary recovery of the very viscous Arlan crude found associated with highly mineralised bed waters.[10–12]

4.4. LATER WORK IN THE UNITED STATES

4.4.1. Union County, Arkansas Field Trial

In 1954 Coty, Updegraff and Yarbrough conducted a field trial in Union County, Arkansas (Yarbrough, personal communication; see also reference 13). Oil was produced from a 30 ft (9 m) layer of Nacatoch sand at a depth of about 2000 ft (610 m). Down-hole temperature was 34 °C, porosity 30%, and permeability ranged from 1000 to 5770 mD. In 1949, when waterflooding began, analysis of core samples revealed residual oil saturations of 4–9% of pore volume. Accordingly, at the time of the trial, the field was considered to be almost completely watered out.

Preliminary laboratory experiments on sandpacks and rock cores were performed with a number of bacterial species. A number of variables were examined but only two, the nature of the core materials and the presence or absence of pressure, appeared to show a reproducible effect on tertiary oil release. A strain of *Clostridium acetobutylicum* was chosen for the field trial on the basis of its ability to grow on sugars in the absence of oxygen and with only simple mineral nutrients to generate pressure by the production of carbon dioxide and hydrogen, and to tolerate temperatures of about 40 °C (104 °F).

The injection water used previously had a salinity of 20–25 000 ppm and, in order to avoid inhibition of bacterial growth, fresh-water injection was started in May 1954. In July 1954, injection of 2% beet molasses into the single injection well was started. This process continued for the five and a half months of the experiment at rates ranging from 100–500 bbl/day (bbl = barrels). Inoculation with the bacterial culture was started in July and continued until November. A total of about 4000 gal (18 180 litres) of dense inoculum was added in 18 separate injections.

The single production well was located about 400 ft (120 m) from the injection well. Samples of gas, water and oil were taken from this and all other producing wells in the vicinity throughout the experimental period. Fresh-water breakthrough at the production well occurred after 70 days and fermentation products (short-chain fatty acids, carbon dioxide, traces of ethanol, butanol and acetone) and sugars appeared at the production well 80–90 days after the first inoculation of bacteria into the injection well. No significant increase in hydrogen content occurred at the production well.

The spectrum of fermentations was not that expected from a pure culture of *Clostridium acetobutylicum* and it was concluded that other bacterial species were also involved in the breakdown of the molasses. Various types of bacteria were isolated from the production well during the trial but the numbers were too low to be correlated with the amounts of fermentation products recovered.

Prior to the experiment, oil production was declining steadily. An estimated decline curve predicted a mean production rate of about 0·6 bbl/day for the period November 1954 to May 1955. The actual daily production rate started to rise in September 1954 and remained well above the projected curve until the end of the experiment in May 1955. Mean oil production during this period was 2·1 bbl/day. None of the surrounding wells showed any increase in production above that predicted, the effects of the treatment were essentially confined to the production well closest to that used for bacterial injection. Microbiologists and engineers connected with the trial agreed that the increased production above that predicted was significant, and was due to the treatment. Unfortunately the mechanism by which the increased yield was produced was not clear. Water production remained fairly constant so the effect was not due simply to a flow increase.

Although the experiment was considered to have been a successful demonstration of a large-scale underground fermentation and enhancement of oil release by bacterial action, no further field trials were carried out. According to Yarbrough (personal communication) this was because the bacterium used was not expected to penetrate formations of low permeability and because at that time there seemed no prospect that the process could be economically viable.

4.4.2.   Other American Studies

The later American activity has included some further field work but this has not been very extensive. In 1961, Bond grew *Desulfovibrio hydrocarbonoclasticus* on a medium containing mineral salts, calcium lactate, calcium ascorbate, yeast extract and 2% agar.[14] The bacteria were inoculated in 5000 gal (22 730 litres) of culture at 900 lb/in^2 (6204 kN/m^2) pressure in a 3000 ft (900 m) deep sandstone well with a 20 ft (6 m) thick oil interval: before inoculation the well produced 15 bbl/day. The well was closed for three months; when production was then restarted the yield rose to 25 bbl/day. Gelatin or carboxymethylcellulose could replace agar as a thickening agent. In low permeability formations the bacteria were infiltrated using fracturing techniques which included the use of sand, gravel, etc., as propping agents. After injection the well was closed for periods of between a month and a year, and production then recommenced. Few production data were reported. In a later patent, Jones found that hydrocarbons with fewer than 10 carbon atoms were oxidised in underground reservoirs by culturing aerobic bacteria in oxygen; cooling water was required to prevent the temperature exceeding 65 °C, the maximum tolerated by the bacteria.[15] Fracturing of the formation assisted bacterial growth. Bacterial slimes could be dissolved by adding hypochlorite and peroxides, and indeed they could be used to form a hot bank of organic material to force through the formation if plugging became serious. A number of bacteria were suitable for inoculation; anaerobes were unable to attack hydrocarbons with fewer than 10 carbon atoms.

A major contribution to later US work has been made by Hitzman. After investigating the ability of a variety of bacteria to release oil from sand, he pointed out some of the problems associated with the injection into formations of live vegetative cells together with nutrients.[16] These included the generation of products in the vicinity of the nutrient (but not necessarily where they might be most useful, close to the oil), plugging due to excessive local growth, and the susceptibility of vegetative organisms to environmental hazards when they are kept metabolically inactive. He suggested overcoming this by injecting spore suspensions, followed later by a slug of nutrient to promote germination.[17] Spores were found to penetrate sand packs more readily than did vegetative cells; they appeared to remain stable *in situ* until germinated. Although this approach might avoid plugging close to the injection face it does seem likely to result in clumping of cells at a multiplicity of germination sites. In a later patent, Hitzman proposed avoiding plugging by drilling the injection well through the oil-bearing zone, and inoculating vegetative bacteria or spores into the neighbouring water-bearing formation; the injection well is then plugged.[18] Growth would be expected to take place at the water/oil interface. An example is given of the injection of 1–1000 bbl of an aqueous suspension of methanogenic organisms (containing more than 10^6 cells/ml) just below the oil zone or at the interface. The water zone is plugged off, and the well completed as a normal producing well. The organisms congregate at the interface and produce methane for 10–30 years, although this has not been checked. The advantages are said to be that any methane produced is an additional benefit at no extra cost; no organisms are produced with the oil; no corrosion or plugging takes place. In a particular instance, for which precise injection conditions are given, the results in terms of oil production are expressed in very vague terms and no proper evaluation can be made from the information in the patent.

Another idea was to use living hydrocarbon-utilising organisms to dislodge and upgrade oil *in situ*. The organisms were to be mixed with crude oil to prepare a slug of material which would be injected into the formation together with surfactant.[19] After a period of time for cell growth the slug would be driven through the formation with saline. No field trials were reported. More recently Hitzman has been concerned with the use of treated bacterial culture medium as a viscosifier; similar work has been carried out by Wegner and Stratton, but neither method involves the injection of bacteria into the formations.[20,21]

A number of conclusions may be drawn both from these studies and from others conducted in parallel by a variety of workers on such related areas as the effects of microorganisms on reservoir fluids, oil–water interactions, compositions of crude oil, and on the physico-chemical and petrophysical consequences of microbial action in the reservoirs.[22-44] It is clear that microbiological activity is possible in reservoir rock at least under some conditions, and that it is capable of influencing the properties of fluids contained therein and some of the mechanical properties of the rock on a micro-scale. Different organisms may attack different components in the crude oil and the balance of these in a consortium will change in response to variation of available substrate. Some of the consequences of microbial degradation may result in lowering the viscosity of the oil, thus facilitating its mobilisation.

In addition to·the possibility of attack on hydrocarbon components, bacteria in the formation may metabolise substrate injected specifically for that purpose: molasses represents the most common material. The bacteria may produce gas, surfactants or acid, all of which can be important for oil mobilisation. Bacterial cells themselves may adhere to the reservoir rock and displace oil. The reservoir conditions are clearly very important in limiting bacterial action: salinity, temperature, pressure, pH and nutrient limitation (inorganic as well as organic) are all important. Finally, if the rock permeability is less than about 100 mD plugging of the formation may occur, though there seems to be little reliable information about the factors determining the likelihood of plugging.

4.5. FIELD TRIALS AND OTHER WORK IN EUROPE

Most of the European activity has taken place in Eastern European countries. However, in one report from the Netherlands aqueous suspensions of several slime-forming bacteria (*Leuconostoc mesenteroides, L. dextranicum, Bacillus polymyxa* and *Clostridium gelatinosum*) were injected into the formation together with the required nutrients.[45] The preferential procedure was to inject the inoculum and nutrient solutions so that the viscosity of the aqueous liquids rose to 500–1000 cP; afterwards conditions were changed so that the injection water was thickened to only 3–5 cP. There was a doubling of the amount of oil recoverable.

4.5.1. Czechoslovakia

In the late 1950s, Dostalek and Spurny injected under pressure into an oil-bearing formation a nutrient solution containing molasses together with a mixed culture of *Desulfovibrio* and *Pseudomonas*.[46] Following an initial increase in the number of bacteria and the amount of produced water, bacterial propagation stopped, water production returned to its original level and oil production increased in some cases by more than 12%. The daily average oil production of the whole formation increased by nearly 7% during the six-month experimental period, the authors regarding this as a consequence of bacterial action. The production of bacterial surfactants was considered the most likely explanation, though this conclusion was not rigorously tested. In laboratory experiments it was found that in sandstone cores the desorption of oil by desulphurising bacteria depended on permeability. In highly permeable cores saturated with oil to 40%, the decrease in saturation by bacterial treatment reached 8%, in less permeable cores saturated to 50–60% the decrease was only 1%.[47] This work appears to have made use of stationary (i.e. non-growing) bacterial cultures. In another study the effect of various nutrients, pH and bacterial mixtures on the ability of sulphate-reducing bacteria to release oil from sterile sands was reported.[48] Strains of *Clostridium* which grow on petroleum and produce large amounts of gas have been isolated from soil; they will also grow in media containing yeast extract and either 2% glucose (or starch) or 4% molasses. These bacteria were used to study de-oiling of sandstone cores, and Dostalek concluded that gas production was the critical factor in releasing oil.[49] His general views on 'secondary' recovery using microorganisms were published in the same conference report though no new ideas appeared there.[50] Dostalek left Czechoslovakia in 1968 and no further work in this field seems to have been carried out in that country.

4.5.2. USSR

Experiments in the Soviet Union reported by Kuznetsov *et al.* have been devoted mostly to the *in situ* generation of gaseous products.[51] Ekzertsev studied the production of methane by bacteria in oil-bearing formations using a mixed bacterial culture.[52] In subsequent work reported by Kuznetsov *et al.* mixed bacterial cultures were inoculated into the reservoir under anaerobic conditions together with molasses; the well was sealed for six months.[51] At the end of this period the oil viscosity had *increased* from 40·3 to 49·3 cP, perhaps indicating a biological utilisation of shorter chain length hydrocarbons. The wellhead pressure had increased by $22·5 \, lb/in^2$ ($155 \, kN/m^2$), the rate of oil production rose from 37 to 40 tonnes/day while the coproduced water decreased by 25%. However, four months later, production had fallen to 36·5 tonnes of oil/day, and the water content had also declined further. The nitrogen content of the gas increased significantly, and that of carbon dioxide and propane increased slightly, but the methane content declined. The total production of gas relative to that of oil rose a little. Clearly this trial was not an economic success, although it does seem to have made use of perhaps the most productive well ever employed in bacterial recovery trials. Except as noted earlier, we know of no further work in the USSR.

4.5.3. Hungary

Work in Hungary, though on a small scale, has been important and took place between 1965 and 1972. The standard technique was to inject the reservoir with a mixed culture of bacteria belonging to the genera *Pseudomonas*, *Clostridium* and *Desulfovibrio* which had been adapted for the purpose. In the earlier work the injection protocol was to introduce $20 \, m^3$ of formation water, followed by a further $100 \, m^3$ containing 4 tonnes of molasses, 120 kg of potassium nitrate, 50 kg of sodium phosphate and 100 kg of sucrose. Next, 100 litres of bacterial culture were added and the procedure was terminated with a final injection of 50 m^3 of formation water. In later trials the quantities of nutrients were reduced to about half those used previously, and the protocol varied to inject part of the nutrients before the bacterial inoculum and the rest after it. Under some circumstances supplementary charges of nutrient would be added at a later date.[53] Although the practice in the initial work was to close the injection well for five to six months and then restart production, apparently on a huff-and-puff basis, it was later recommended that the injection well should remain closed, except for periodic additions of nutrient, so long as effects were observed in neighbouring production wells.[53,54] The well spacings were not specified. The use of supplementary additions of nutrient was indicated by an increase in the proportion of produced water.[53] Some years later the prescription was altered so that when supplementary injections were being used these should be halted after two to three years to allow the products of bacterial metabolism to be eliminated, while continuing to produce oil from the production wells.[55] The significance of this is not clear to the present authors.

While a number of interesting observations were recorded very few quantitative data on oil production were given. In one experiment involving a gritstone formation at depths of more than 600 m, temperature of 50 °C and a permeability in the range 600–700 mD, an increase in the rate of oil production was observed over an eight month period. The viscosity fell from 42 to 18–26 cSt (40 °C), the produced water was neutral rather than alkaline, and the produced gases (11% carbon dioxide) increased from zero to 40 m^3/day. The bacterial action was said to affect an area of 60 000 m^2. However, no effects were observed in another gritstone reservoir at a depth of 1400 m, with a temperature of 50 °C and a permeability of 10–70 mD.

Experiments were also carried out in calcareous rocks at 700 m and 50 °C. Increases in produced oil of up to 60% were recorded, lasting between 2 weeks and 18 months in different wells. As before, oil viscosity and the pH of the produced water fell and the proportion of carbon dioxide in the produced gas increased.

The deepest reservoir employed was one at a depth of nearly 2500 m, in which the pressure was 228 atm and the temperature 97 °C. The well spacings from the injection to the production wells were 300–1700 m. It was observed that bacteria, particularly *Desulfovibrio*, spread through the formation, and the numbers of organisms increased; clearly growth had taken place, even at that very high temperature. It was further observed that the natural bacterial flora of the reservoir was stimulated to grow by the addition of bacteria obtained from other sources such as sewage and well-bottom muds. This was interpreted as meaning that the added bacteria metabolise the added molasses to products which could be used as nutrients for the indigenous organisms. However, the present authors are once again puzzled, this time by the term 'indigenous organism'. It implies that virgin reservoirs contain a natural population even at depths of nearly 2500 m. We are not sure that this meaning was intended.

4.5.4. Poland

All of the Polish work has been published under the sole authorship of Karaskiewicz at the Petroleum Institute in Krakow. He undertook an extensive programme of work between 1961 and 1971 in which 20 wells in sand reservoirs at depths from 500 to 1200 m were submitted to microbiological treatment. A number of papers and conference reports have been published and many of his experimental data are reported in considerable detail in a monograph.[56–62] He obtained his bacterial strains (*Arthrobacter, Clostridium, Mycobacterium, Peptococcus* and *Pseudomonas*) from soil, crude oil, formation water and industrial wastes, particularly from the neighbourhood of sugar refineries. The strains were adapted to oil mobilisation in an apparatus called a 'collector' modified from an earlier design by Jaranyi *et al.*[57,63,64] The typical pattern was to inject 500 litres of mixed bacterial culture (containing 3–6×10^6 organisms/ml) together with 2 tonnes of molasses and 50 m^3 of produced water from the well being injected. A slug of 150 litres of oil was added to seal the well which was reopened after six months and production started again. All operations were on a huff-and-puff basis with oil being produced via lift pumps, although effects were observed in neighbouring wells. While the well spacings were not given it is likely that wells were close together. Extensive data were reported for oil, gas and water production, gas composition, numbers of bacteria and some physical properties of the oil. The data for gas yield were not very reliable but those for water and oil were good. Individual wells discharged into collecting tanks which were monitored twice a day; water and oil were separated in separators at the wellhead. The operating practice was for the proportion of oil to water being produced to be inspected visually at least once during each 12 h shift period. If oil was being produced, pumping was continued. If water was being produced the pump was shut down until the beginning of the next shift. Thus, a well yielding a high oil:water ratio would be pumped for more hours per day than one with a poor ratio. Before injection with bacteria, well production was logged for a year or more to establish a base line for extrapolation during the period of microbiological treatment. It should be noted that the wells in question were producing a few barrels of oil per day.

The average extent of enhancement of recovery was in the range 20–200% for different wells for periods lasting between two and eight years. Very often (but not invariably) the period of well closure was followed by one of several months at least during which oil production was particularly high and in itself often was enough to compensate for the lost production during the time the well was not producing. Production then settled down over a long period at a steady enhanced rate which was nevertheless below the peak value. In some instances hydraulic fracturing was employed as much as three years after the initial injection. With wells in which the enhanced rate of production had not diminished it was virtually doubled as a result of fracturing. In at least one case a well which had shown no response in its oil production to the injection of bacteria underwent a three fold increase when fractured, though it would be difficult to demonstrate in that well that the bacteria were relevant and that the response was not due solely to fracturing. As the Hungarian experience had shown earlier, it may be advantageous to introduce a second charge of nutrient after an appropriate period.

Karaskiewicz is no longer working on bacterial enhanced recovery methods and hesitates to draw firm economic conclusions from his work. He had access only to depleted fields and so may have been working under unfavourable circumstances. The cost of an injection into a well using a pump which was to hand, and ignoring the cost of the inoculum, was about 9500 zl; this was equivalent to the purchase price of about 5 tonnes of oil. If rock absorption was high a larger pump had to be hired and this raised the treatment cost. The costs included in the zloty figure comprise mostly the hire costs for a pump and a truck, the price of the molasses and other transport costs. The wages of people normally employed do not seem to have been charged. The costs of fracturing operations were additional and very much higher than the basic figure indicated above. Limited though this analysis may be, it is the only one known to us.

In choosing injection wells a number of considerations were judged important including tectonics, structure and petrophysical properties of the formation, chemical composition of the oil, temperature, pressure, etc.[62] Care was taken to choose wells in a good state of repair with casing in good condition to the bottom. The increased yield of oil was correlated in a number of cases with the size of the bacterial population present. Bacteria were observed to migrate through the formation from the point of injection since they appeared at production wells some distance away; no rates of migration are given and either the time when bacteria appeared in a neighbouring well, or its distance from the injected well, is omitted. It is stated that continuing bacterial activity underground depends upon the supply of carbohydrate nutrient, although elsewhere it is stated that some attack on the hydrocarbons takes place when the molasses is exhausted. In Karaskiewicz's view the production of gas and surfactants by the bacteria in the formation is the most important factor leading to enhanced recovery in his trials. Analysis of the produced gas typically showed a high content of hydrocarbons, presumably derived from bacterial action. In one well in which no gas had been obtained for several years, the gas pressure increased to 8 atm 110 days after inoculation with bacteria. The composition of this gas was methane (89·6%), ethane (6·1%), propane (2·8%) and butane (0·5%); these percentages are of the total hydrocarbons; the proportion of non-hydrocarbon gas was not given. In another case the overall gas composition several years after inoculation was air (2·6%), methane (81·0%), ethane (0·9%), propane (2·22%), isobutane (2·15%), n-butane (2·5%) and carbon dioxide (8·63%). Karaskiewicz's general conclusions are that microbiological methods are cheap, easy to apply, require little special equipment and do not destroy the reservoir. In this he was supported by Hitzman, who pointed out (as did Dunlop on another occasion) that more work needs to be done and surmised that microbiologists may yet recover oil where engineers have failed.[65,66]

4.5.5. Romania

Work along the lines pioneered by Dostalek, Jaranyi and Karaskiewicz and directed towards the fields at Ploesti was started in 1971 by a group under Lazar at the Central Biological Institute in Bucuresti, in collaboration with reservoir and production engineers. A recent summary of these activities was presented as a conference report.[67] The investigations have proceeded through three stages:[68] characterisation of the bacteriological population in the formation waters of suitable reservoirs; adaptation to reservoir conditions of bacteria selected primarily on the basis of their high metabolic activity, and their abilities both to mobilise oil in laboratory experiments and to utilise crude oil; field trials.

A variety of bacterial forms were found in formation waters.[69-72] In hot, deep reservoirs cocci and spore-forming bacilli were the most common types; in shallower formations non-sporogenic bacilli, particularly motile pseudomonads, and members of the enteric bacteria predominated. The richest ecological environments for obtaining bacteria able to ferment molasses in the presence of formation water and oil were the 'waste muds' from sugar refineries, formation water itself, well mud and the sediment from the plant used to purify injection waters.

Adaptation procedures followed basically the Hungarian and Polish procedures,[62] with local improvements and developments.[73,74] Using either pure or mixed cultures not previously adapted to oil mobilisation a release of 14·5% of the oil in place was obtained in laboratory experiments, rising to 32·5% for adapted cultures.[75] Field work therefore always employed adapted mixed natural populations. The adapted populations contained mainly representatives of the genera *Pseudomonas, Escherichia, Aerobacter, Arthrobacter, Mycobacterium, Micrococcus, Peptococcus, Bacillus* and *Clostridium*. Smaller numbers of individuals belonged to *Desulfovibrio, Cellulomonas, Brevibacterium* and *Flavobacterium*, and there were some yeasts. Laboratory experiments showed that the adapted bacterial populations released as much as 40–45% of the oil-in-place with the initial charge of nutrients. Supplying additional quantities of nutrients raised this to 50%, and sometimes to as much as 60%. These yields were influenced by the proportion of sand to oil used in the model experiments (4:1 was the most productive ratio) as well as by the proportion of nutrient medium containing molasses to bacterial inoculum (volume ratios in the range 6:1–19:1 were best). The inclusion of sugar refinery mud led to a still greater yield, then reaching 70% or more. In those experiments it was found that, under the different conditions, the bacteria consumed between 1% and 35% of the oil. Karaskiewicz had earlier concluded that under some conditions the bacteria would metabolise oil when the molasses was exhausted, as we have noted above.[62]

In the period 1976–78, nine wells in Romania were subjected to microbial enhancement. These were located in sand reservoirs, at depths of 500–1550 m, with permeabilities of 80–1000 mD and temperatures in the range 27–56 °C. The coproduced water was between 30% and 100%, with sodium chloride concentrations from 5000 to 200 000 ppm.

Three types of wells have been injected: those still producing oil (0·2–1·0 tonne/day), those producing only water, and those employed as injection wells for waterflooding. The protocols for the injection of producing wells followed the Hungarian methods,[27,54,55] while the technique proposed by Hitzman was used for the waterflooding of injection wells.[19]

Little information is so far available on the results of these activities. In his most recent paper, Lazar reports only that encouraging results have been obtained. These apparently include changes in the physico-chemical properties of the produced liquids and some 'significant increases' in oil production. Although no details have been released it seems likely that the sort of increases announced by Karaskiewicz have also been obtained in Romania.[62] The work of Lazar's group continues and may at the present time be the only remaining European activity in this area.

Like Karaskiewicz, Lazar has worked only with partly or completely exhausted wells. Even his most fruitful wells are described as 'those that still produced oil.' This is another instance where the full potential value of the method is difficult to evaluate because of the poor quality of the wells employed and the total absence of data on which to base a cost-effectiveness analysis. We shall return to this problem.

4.6. AVAILABILITY OF WELLS FOR MICROBIOLOGICAL FIELD TRIALS

With perhaps one or two exceptions, it is certainly true for all the Eastern European field trials, and probably also for the American ones, that the injection of bacteria has been permitted only into wells which always were poor producers or were near the end of their productive lives. It is not surprising that this should have been the case. It is normally the practice in oil production not to employ enhanced recovery techniques before it becomes necessary, i.e. before the productivity of the well is seen to be falling to the point where it might be better to close it altogether. Thus, neither waterflooding, nor pressure maintenance, is put into use so long as the primary mechanisms of gas cap drive, gas solution drive or water drive give adequate rates of oil production. Only as the yields from these drive mechanisms fall are secondary techniques brought into play; tertiary methods (surfactant and micellar slugs, carbon dioxide flooding, etc.) tend to be employed even later, if at all. Microbiological methods, too, have always been investigated very late in the production life of the well.

4.6.1.   Assessing the Benefits

One obvious consequence of these practices is the difficulty of assessing the benefits of tertiary methods. Whereas the protagonist of the method observes, perhaps, a large *percentage* increase in the rate of oil production, the sceptic perceives a relatively trivial addition to the yield expressed in absolute terms. Thus, in the case of the work in Poland, increases in yield of more than 100% were observed in some instances, lasting often for years on end. In absolute terms, however, the increase was often from about 1 bbl/day to about 2 bbl/day; even at a world price of $35/bbl this cannot be seen as economically very exciting.

4.6.2.   Early or Late Injection of Microorganisms?

Would there be a greater proportional and absolute benefit if bacterial treatment were started relatively early in the life of the well, perhaps at the end of the primary drive phase? Since as far as we know this has never been attempted we are unable to give a firm answer. And until experience has been gained such a practice is hardly likely to become standard operating procedure. So one aspect of making further progress in this direction depends on a willingness to allow early bacterial enhancement procedures on an experimental basis. Such a willingness must imply a recognition of the possibility not only that the experiment itself might be unsuccessful in achieving the desired end (that is a risk inherent in any experiment) but further that damage might in some way be done to the well which would reduce the cumulative production benefits below what might have been expected without these procedures. There is a further uncertainty. The production profile of a new well cannot be predicted with confidence. The use of bacterial (or other enhanced recovery) methods on a well with little or no production history makes it very difficult subsequently to decide whether or not enhancement has actually occurred. Would not that production yield have been obtained anyway? Might it not have been even greater?

The arguments in favour of the early use of bacterial enhancement methods are related mostly to the ease with which the bacteria will reach difficult parts of the reservoir. In a waterflooding procedure, in which bacteria were employed when the produced oil/water ratio was already falling seriously, it would seem likely that the bacteria would tend to move predominantly along established flow channels from which the oil would already have been expelled. If bacteria were successfully deployed at the very beginning of the waterflood it is more probable that the degree of recovery from the totality of the swept area (including those lenses and ganglia not actually part of the main flow stream) would be higher, and that a greater net yield would be obtained before the production well became watered out. The reasons for these conclusions are that those bacterial products promoting oil mobilisation would be produced on a broader front and from an earlier time than without bacterial treatment, so that breaking through, fingering and bypassing would be expected to be suppressed.

We surmise that similar considerations would apply on a huff-and-puff basis. Again, with early application the bacteria would come into closer and more uniform contact with the bulk of the oil in the neighbourhood of the well and their restriction to predominantly aqueous channels would be less pronounced. If bacterial methods were to involve the simultaneous injection of molasses or other nutrient, the common practice in most of the field trials so far attempted, then the efficacy of bacterial action depends upon both the cells and the nutrient being available at the same place. So even if bacteria were able to migrate out of the main flow stream into unswept areas by virtue of their own motility, the nutrient materials would still be carried on the flood and bacterial action would soon cease for lack of food and fuel.

Resolution of the problem of how to test the action of bacteria on new wells is not made easier by the shortening time scale. There now exist many wells throughout the world that are already non-productive and thus, at best, candidates only for terminal enhanced recovery. The rate of discovery and bringing on stream of new wells has already, or will shortly, decline from the peak. The longer the delay before bacterial methods are perfected, the greater the number of wells that will become partially or fully exhausted, and the fewer that remain as candidates for early treatment. If indeed early bacterial treatment would produce, as we suspect, a greater net yield than late bacterial application, we are steadily losing opportunities for its most successful deployment and equally losing much of the potential reward. It seems to us, therefore, a matter of urgency that renewed development of bacterial methods should be initiated without delay, and there are signs that this is beginning to happen. New techniques and understanding in biological manipulation have opened up possibilities for obtaining suitable microbes to a degree hardly dreamed of when most of the bacteriological enhanced recovery field trials were being conducted. With probably more than 70% of the world's discovered oil-in-place not currently recoverable the opportunities are clearly very great.

Many reservoirs world-wide are so deep or in such relatively inaccessible offshore locations that the strategy for their working is very different from that which is economically and physically possible with shallow onshore formations. We shall discuss the more difficult microbiological problems of the deep or inaccessible reservoirs in the next chapter. In the present chapter we shall explore further, on the basis of experience already gained, the techniques, possibilities and limitations with shallow reservoirs.

4.7.   CHOICE OF SUITABLE WELLS

Most of the authors with extensive experience of field trials refer to the need to choose suitable wells. Although some discussion has appeared of relevant well properties all of the authors are vague about actually making choices and how they might view weighing up one favourable (or unfavourable) characteristic versus another. It seems likely, particularly in view of our discussion in the preceding section, that the choice of wells was not made predominantly on microbiological grounds but was heavily influenced by management and production considerations.

In Karaskiewicz's view the intensity of microbial activity in the reservoir is influenced by the chemical composition of the bed, by its petrophysical properties and by a number of purely biological factors.[62] He regards the most important considerations as being the mineralisation of the bed waters, the availability of bacterial nutrition and the temperature. The water mineralisation may have a critical role in bacterial development; water with high concentrations of sodium chloride, or of magnesium ions, would have a pronounced toxic effect on most microorganisms. Relatively few types of bacteria are able to withstand the inhibitory effects of high salt concentrations, and under such conditions it would clearly be necessary to use halotolerant forms. An unfavourable pH, as well as shortages of nitrogen and phosphorus in forms which can be readily assimilated are all further possible limitations. Both high temperature and high pressure are potentially deleterious factors. As we have discussed in an earlier chapter, the ability of bacteria to move through the formation fairly readily, and neither to be confined close to the point of injection, nor to cause plugging of the rock channels, is a major determinant of the suitability of a particular reservoir for bacterial treatment. This is a topic about which there is not yet sufficient information to permit a definition of a suitable minimum permeability of formation rock, at least as formulated in conventional terms. It seems clear that rock permeabilities of 100 mD or more will present little difficulty, but there must be a minimum degree of penetration for a bacterial method to be feasible. The finding by Myers and McCready that bacteria can penetrate rock with permeabilities as low as 0·1 mD is difficult to interpret:[76] do bacteria penetrate through very narrow channels, or is such a relatively low permeability compatible with a matrix of many narrow channels accompanied by a few large enough for bacterial penetration? If the latter, would such a system be amenable to bacterial treatment? Even if the bacterial cells were unable to reach every nook and cranny in the reservoir rock, the release into solution of effective bacterial products in their immediate neighbourhood may itself help significantly to mobilise oil.

Very little has been reported about the effects on bacterial enhancement of recovery of the degree of saturation of the rock by oil, although Dostalek has shown that microbiological processes are most effective at lower levels of saturation;[49] at a saturation of 25% about 20% of the oil was released in experiments with drill cores, whereas at 60% saturation the release was as low as 10%.

Lazar also gives little information.[67] He notes that three types of wells were used for injection (some producing only water, some used for waterflooding and some still [sic] producing oil) but he does not discuss the criteria on which the choices were based.

4.8. ISOLATION OF SUITABLE BACTERIA

While some of the earlier workers both in America and in Europe used pure strains of bacteria for their laboratory experiments and field trials, the Hungarian, Polish and Romanian procedures have made use of *ad hoc* mixed cultures, the components of which were later identified.

4.8.1. Adaptation to Reservoir Conditions

Karaskiewicz has given most information about the methods of obtaining mixed cultures 'adapted' to reservoir conditions. No mutagenesis or specific selectional techniques were used; rather bacterial populations from a number of sources were subjected to a simple selection for their ability to mobilise oil under conditions somewhat resembling those of the reservoir. He used either a simple inorganic medium, or one supplemented with 3% molasses, or one containing 4% molasses and 0·1% natural crude oil filtered through a Seitz filter together with bed water from the formation from which the particular bacterial population was to be isolated. In some cases the bacteria were grown, probably anaerobically, in full stoppered bottles; in others the medium was solidified with agar, but whether or not these were anaerobic is not stated.[62]

4.8.2. Use of a 'Collector'

Some adaptation experiments were carried out in a model system described as a 'collector', described most fully in a conference report.[57] This consisted essentially of a vertical tube, 10 cm diameter and 200 cm long, packed with river sand saturated with oil.

An inlet tube at the bottom permitted the infiltration of the bacterial culture, while another at the top led into the side of a flask to collect the displaced oil. From the top of the flask another tube led into the top of a second flask; from the bottom of the latter an outlet tube led into an overflow vessel. Thus, oil liberated from the collector tube would collect in the first flask, produced gas in the second, while the displaced brine originally filling the two flasks would collect in the overflow vessel. The bacterial suspension was admitted to the collector under a hydrostatic pressure of 0·15 atm. The whole apparatus was maintained at around 25 °C in the dark. The medium consisted of natural formation water + 4% molasses (pH 8·9) and some 500 ml of suspension containing 8×10^5 organisms/ml were injected.

After 5–8 days gas bubbles were formed and the migration of oil was observed after 15–18 days; distinct dark spots were seen in the sand which later migrated upwards.

In one reported experiment (and another was rather similar) 13 950 g of sand grains (0·5–0·6 mm diameter) were placed in the tube together with 1537 g of black paraffinic crude from a Carpathian Eocene sandstone stratum (specific gravity at 20 °C, 0·842; viscosity at 20 °C, 7·13 cP). After about six months 837 g (some 54%) of the oil had been mobilised, 5600 ml of gas released and the pH of the aqueous phase had fallen to 5·6. Analysis of the liberated gas showed a predominance of n-butane, isopentane and n-pentane in the first few months, but after more than a year methane formed nearly 40% of the total with no hydrocarbon of greater molecular weight than butane being present.

4.8.3. Alternative Forms of Apparatus

A second type of apparatus used drill cores from various limestone and sandstone strata.[62] The cores (of unstated dimension) were heat dried and then saturated with oil by vacuum infiltration. Each was placed in a 2 litre bottle completely filled with a 4% aqueous molasses solution and inoculated with the mixed bacterial culture. A tube led from the top of this flask to the top of another completely filled with brine. As before, a further tube led from the bottom of the second flask to an overflow vessel. During incubation for six months at 32 °C checks were made of the bacterial flora, pH and liberated gas.

For preparing larger quantities of bacteria for injection into wells 10 litre flasks were used containing a 10 cm thick layer of sand mixed with bacteria and oil. The flasks were filled with formation water containing molasses. A system of interconnecting tubes and overflows ensured that the gas resulting from bacterial action displaced part of the culture into a second flask where the pH was readjusted and further nutrients added. The medium was then returned to the original vessel and recycled until a population of 3×10^6 bacteria/ml was obtained. This culture was then used for injection in the field. Diagrams of all three pieces of equipment appear in reference 62.

4.8.4. Further Modifications

Lazar, again only in a conference report,[77] described further modifications of these techniques and compared the results he obtained with his own version of Karaskiewicz's third apparatus[62] with his modified pieces of equipment, one of which contained an agitator. All of these modified designs gave populations better able to mobilise oil than the Polish design, but the best response was obtained by combining all four mixed cultures.

4.8.5. Evaluation of Adaptation Procedures

Both Karaskiewicz and Lazar used the same primary sources of bacteria: bed waters, well muds, the surface soil from around the wells, sediment from the plants used to purify formation waters, and effluent waste waters from sugar refineries. What they have done essentially is to set up laboratory models crudely reflecting some of the reservoir conditions which the bacteria would meet when inoculated into the oil-bearing formation. These included the use of bed waters and crude oil from the formation to be inoculated, a temperature corresponding to that in the formation to be used, and in some cases an absence of oxygen. No attempts were made to use high pressures in laboratory experiments. Since it was a part of their procedure to inject molasses into the reservoir together with the bacteria, this material was also present in the model systems. Both workers have then allowed populations to develop from their various sources which can survive in the laboratory model (and hence presumably in the actual reservoirs themselves) and which have demonstrated some ability in the release of oil. Very prolonged laboratory experiments have not been attempted to assess the stability of the biological systems, although it is clear that some, at least, of Karaskiewicz's laboratory studies lasted for more than a year.[57] The important evaluation of this work, however, lies not in how the system behaved in the laboratory but how effective it was in field trials. Its success in Karaskiewicz's hands is clearly documented,[62] and while Lazar gives little detail he, too, claims to have obtained encouraging results.[67]

4.9. EVALUATING THE BENEFITS OF ENHANCED RECOVERY PROCEDURES

Because no two wells are ever identical it is not possible to mount properly controlled experiments except on a statistical basis. Thus, a number of wells as similar as possible could be identified, some of them inoculated with bacteria and others not treated but injected with liquid not containing bacteria to serve as controls. (This might be rather unsatisfactory as the injection even of sterile medium would encourage contamination.) The wells would, of course, have to be so chosen to ensure that organisms injected into the experimental wells were not able to influence the controls. No experiments as extensive as this have been conducted.

In the absence of large-scale trials the evaluation of an enhanced recovery procedure would have to depend on an extrapolation of a well's historical performance. This in turn depends on the accuracy and completeness of logging in the past; its absence would be one of the factors limiting the application of enhanced recovery procedures early in the life of a well. In his monograph Karaskiewicz showed rather satisfactory logging data for oil and water production for periods of one to several years prior to the injection of bacteria.[62] The depletion curves were extrapolated forward for as much as eight years, and it was against that extrapolation that enhanced recovery was judged. The extrapolations were done conservatively and there seems little reason to fear that unexpected discontinuities would have seriously invalidated the conclusions.

In other low-yielding shallow fields, for which the injection of adapted bacteria plus molasses might be economically attractive, the logging data are less satisfactory. The present authors are aware of cases in which logging of individual wells has not been carried out so that no proper baselines are available. Unless the enhanced production is so large as to be unequivocal it may be difficult to recognise it convincingly even though it is taking place. In another example involving an attempted waterflood known to the authors the volume of water injected was considerably greater than the combined volumes of oil and water produced. Clearly, unless reliable data are to hand no proper evaluation can be made.

4.10. LIMITATIONS OF THE METHOD AND THE RISKS OF FAILURE

As with any industrial process, there are limits beyond which the bacterial enhancement method is unlikely to be effective even under the most favourable of circumstances. There is also some possibility of failure in any individual enhanced recovery application and this, too, is a universal attendant of manufacturing activity. The frequency of a successful outcome is probably less in any case than for an established routine industrial procedure since enhancement techniques applied to different wells must always operate in at least somewhat different circumstances; enhanced recovery cannot be as routine an operation as the manufacture of a product on an assembly line. Among the potential problem areas are included the following:

4.10.1. Plugging the Formation
The risk of plugging the formation by bacteria and their products

during water injection is well known.[78,79] The relationship between the tendency to plug and the parameters of formation rock is, however, not properly understood. Neither permeability nor irreducible water saturation of the rock was simply related to plugging susceptibility, nor were various combinations of rock properties.[80] Bacterial plugging of the sand face of water injection wells appears to be of two types: plugging by the organisms themselves (including iron bacteria, slime formers and a large variety of bacteria, algae and fungi), and plugging by bacterial products (such as the precipitates of metallic sulphides following the formation of hydrogen sulphide by sulphate-reducing bacteria).[81] The obstructive effects of the bacterial cells will depend on many factors and may be due both to an individual cell becoming entrapped in a narrow orifice (or partly entrapped, so as to constrict the natural orifice further and promote plugging by another cell) and by colonies of organisms growing perhaps in a rather slow-moving water stream. In spite of the application of bactericides it must be virtually impossible under operating conditions to keep an injected reservoir truly sterile. The risk of plugging does then depend very much on the suitability of the reservoir for bacterial growth. An example of this is provided by a waterflood which operated as a closed system with deep-well brine as make-up water. Sulphate-reducing bacteria were present but did not become sufficiently prolific to cause significant plugging or corrosion. This contamination with sulphate-reducing bacteria is known to have continued for 11 years without plugging problems.[82]

The cell density in the bacterial inoculum is probably also important. In his inocula Karaskiewicz used cell densities of between 0·9 and 10×10^6/ml;[83] Lazar implies the use of inocula containing of the order of 10^9 cells/ml.[77] Neither has reported serious problems of plugging. Few experimental data are available on the effects of plugging by different cell concentrations injected into formations of different permeabilities. One report does suggest that while rock with a permeability in the Darcy range will not plug with 10^6 cells/ml, a reservoir showing 20 mD would plug with 3×10^4 bacteria/ml.[84] The same author also states that bacteria-free injection water is required at 1–2 mD, but this is clearly at variance with findings of bacterial mobility through rock with a permeability of less than 0·1 mD.[76]

4.10.2.   Unfavourable Geological Conditions

A counterpart of the plugging risk is the ability of the injected bacteria to move through the formation. Anything which stops or hinders such movement is expected to reduce the efficiency of treatment. A low permeability promoting plugging will clearly not be conducive to penetration. Geological faulting, or occlusion of pockets of crude behind impenetrable rock barriers, will similarly reduce potential invasion of the formation by bacteria of the enhanced recovery procedure.

Under this subheading we formally include other aspects of reservoir conditions (temperature, pressure, salinity and other mineralisation of bed waters or injected waters) but these will be treated more extensively in Chapter 5.

4.10.3.   Unfavourable Characteristics of the Crude Oil

Some crudes will be more susceptible than others to mobilisation by the products of bacterial metabolism. Generally lighter crudes may be expected to respond more readily than heavy oils. Even though bacterial action may promote mobilisation in many grades of oil, the extent of promotion is likely to be insufficient with heavy or asphaltic crude oils. The limits of bacterial efficacy do not seem to have been studied on a formal basis, though some relevant information can be inferred from studies of non-biological chemical floods.

4.10.4.   Contamination by Deleterious Microorganisms

The injection of the chosen culture of bacteria can be carried out with a minimum of contamination by unwanted forms. Thus, the injection waters may be sterilised (whether by a chemical or physical method would depend on the details of the local operation conditions) and after removal or inactivation of the sterilisant the active culture added. Under field conditions, however, one cannot expect always to achieve absolute success in eliminating undesirable bacteria. They are virtually ubiquitous in their distribution, very small indeed, and that ability to propagate which means that only a small inoculum of active bacteria need be delivered to the neighbourhood of oil in the reservoir also means that a very small number of contaminating organisms might be very damaging under certain circumstances. The risk is best avoided by using active bacteria with such properties, in a nutritive medium to which they are so well adapted, that in competition with undesirable contaminants in the reservoir the preferred forms will be the more biologically successful and will dominate the contaminants. This approach requires a proper understanding of the nature of a possible biological competition. As with all procedures there must always be a residual risk of failure, but it might be made very small indeed. The risk would arise if the unwanted contaminants mutated to new forms, better able to compete in the reservoir, but retaining their undesirable properties. This risk can be minimised by presterilisation of the injection waters, since even if competitive contaminants did enter the system they would do so in such small numbers as to be overwhelmed by the active culture. To compete successfully under such conditions the contaminants would need an enhanced level of adaptation to the reservoir which is most unlikely indeed to occur. Generally, we may conclude that even if presterilisation is not absolutely successful in keeping contaminants out, it is an important means of reducing their ability to cause harm.

4.10.5. Insufficient Nutrition

The injection of bacteria into the formation together with a supply of nutrients will permit bacterial action only until the nutrients are exhausted unless in the reservoir the bacteria are able to make nutritional use of some of the crude oil components. It has been shown by Karaskiewicz that in treated wells in which enhanced recovery was declining it was possible to stimulate it by the injection of more nutrient (molasses) even though the bacteria were said to be able to subsist on hydrocarbon fractions when the carbohydrate had been exhausted.[62] It should be remembered that he used bacteria selected for their ability to mobilise oil in the presence of molasses.

Since most, if not all, of the bacterial well treatments have been done on a huff-and-puff basis we may conclude that the effects of the bacteria were fairly local in the immediate neighbourhoods of the injected wells. A second injection of molasses might then reach that local area fairly well. Whether it would do so quite so readily in a waterflood situation is less certain; with that type of approach only nutrients injected more or less together with the bacteria might be of any value in sustaining them. Because there has been little if any experience of using the molasses technique with a waterflood one must reserve judgement as to whether or not it is a feasible approach to enhancing oil recovery. As we shall discuss at greater length in the context of the deep reservoirs with wide well spacings, it might, under those circumstances, be possible to combine a waterflood with bacteria able to subsist on the hydrocarbons of the crude oil.

4.10.6. Collapse of the Biological System

Any process must have some probability of failure in use, and one must admit that however well designed and constructed a microbiological system may be it, too, will sometimes behave in an unexpected way. Once more there is not sufficient data to decide whether the risk of this happening is very great as an evaluation must depend primarily on the employment of the system in the field. Again we must rely mainly on Karaskiewicz: in some of his trials the enhancement of recovery went on for at least eight years after bacterial injection, and was still continuing at the time his reports terminate. In some of his trials the response was minimal and it is not easy to analyse the reasons for a poor result. Perhaps the nature of the reservoir conditions was misjudged and they were less favourable than had been hoped. Perhaps, indeed, the trial had been made on an experimental basis, even though the reservoir conditions were known to be poor, in order to gain experience. Or again, perhaps the bacteria simply failed to act in the expected manner, and short of sinking test bores into the formation, there was no good way of finding out why the failure occurred.

Without knowing the reasons for failure it is difficult to predict the consequences. If it were purely that the bacteria failed to survive no damage might have been caused to the reservoir, though expense may well have been entailed to maintain the operation in the expectation of a response which was not forthcoming. That sort of risk is implicit in any enhanced recovery operation. If, on the other hand, the bacteria unexpectedly caused extensive plugging of the formation, or were successfully displaced by contaminants which had undesirable effects on the reservoir, or its installations or on the produced oil itself, then actual damage would have been suffered over and above the failure to acquire the additional recovery of oil. Only further experience will resolve these problems and quantify the risks.

4.11. OPPORTUNITIES FOR SUCCESS WITH BACTERIAL ENHANCEMENT METHODS

The trials already undertaken have yielded results strongly suggesting that bacterial enhancement of oil recovery is a real possibility at least in some cases. Assuming, then, that the molasses-based technique, within the limitations discussed in the last section, were to form the basis of an operational procedure we can consider in general terms what investments would be required and what returns conceivable.

4.11.1. Research and Development
Sufficient information is already available to permit fairly rapidly the isolation and adaptation of bacterial populations for injection into wells together with molasses, once particular wells have been chosen. A reasonable estimate in terms of effort might be perhaps six man-years of work by suitably skilled personnel to institute the general techniques and obtain adapted populations for a small number of identified wells. Thereafter, populations adapted to additional wells might be obtained at the rate of several per man-year on the assumption that no difficult and unexpected problems are encountered.

One of the most important missing pieces of information at the present time is the detailed relationship between rock permeability and bacterial penetration. We perceive the results of such a study as being an essential benchmark against which to judge the suitability of particular strata for treatment with bacteria. Thus, unless this information becomes available from other sources, we think it will be incumbent on an organisation intending to develop techniques in the bacterial enhanced recovery field to obtain the information for itself. As a rough estimate we think that it would take several man-years of experiment to do so.

4.11.2. Cost of Injection

Once a workable system had been acquired, and experience gained on how best to use it, the marginal costs attendant on injecting each successive well should be roughly calculable. Karaskiewicz's cost as the price of 5 tonnes of crude oil (say, around $1250 at April 1981 prices) seems reasonable for materials, transport, etc.; to this one should add a fairly small proportion (say, 10–20%) of a microbiologist's cost of employment, including salary, fringe benefits and institutional overheads.

The injection having been made, the running costs of maintaining production should not rise appreciably just because more oil is being produced from a well as the result of a successful operation.

4.11.3. Benefits from an Increased Yield of Oil

Some of the Polish wells showed a doubling of the daily rate of oil production which persisted over a period of eight years and was continuing unabated when the report was published. Even at the rate of doubling the very low production of 1 bbl/day to 2 bbl/day, a cumulative increase of over 2900 bbl would be obtained in eight years, worth more than $105 000 at the April 1981 price of about $35/bbl. Many wells, of course, produce much more than 1 bbl/day but we do not know whether a well producing at many times that rate could also be stimulated to double its output.

Furthermore, in all that we have written about the success of certain enhanced recovery trials with bacteria we have tacitly assumed that an enhanced rate of recovery, especially over a period of several years, implies a greater net recovery of the oil-in-place. Such a conclusion is really an extrapolation from laboratory model experiments in which the quantity of the oil-in-place is accurately known, and the residual oil-in-place after no more is mobilised can be measured with confidence. In the field, however, it is possible to reach such conclusions only over long periods of time and by a statistical comparison of the cumulative yields from treated and untreated wells. But even if the ultimate *yield* of the oil-in-place were not raised, an important economic benefit would be derived from an increased *rate* of production of that proportion of the original oil which will eventually be produced. Since operating costs are fairly independent of the rate of oil production, it is economically advantageous to produce a given quantity of oil in a shorter rather than a longer period.

The practice commonly used of closing a production well for six months or so following the injection of bacteria necessarily entails a loss of production revenue during that period even while the overhead operating costs continue. So a period of enhanced oil *production* is required simply to break even financially and before any net benefit is obtained. It is interesting in this context that the period of well closure is often followed by a period of a few months of particularly rapid oil production, which more or less balances the loss. Thus, under favourable circumstances a net financial gain consequent on bacterial injection may begin to accumulate in as little as a year after inoculation.

REFERENCES

1. Beckman, J. W. (1926). *Ind. Eng. Chem., News Ed.*, **4** (10 Nov. 1926), 3.
2. ZoBell, C. E. (1946). US 2,413,278.
3. ZoBell, C. E. (1947). *Oil and Gas Journal*, **46** (13), 62.
4. ZoBell, C. E. (1947). *World Oil*, **126** (13), 36; **127** (1), 35.
5. Updegraff, D. M. and Wren, G. B. (1953). US 2,660,550.
6. Updegraff, D. M. and Wren, G. B. (1954). *Appl. Microbiol.*, **2**, 309.
7. Updegraff, D. M. (1957). US 2,807,570.
8. Beck, J. V. (1947). *Producers' Monthly*, **11** (11), 13.
9. Kuznetsov, S. I. (1950). *Mikrobiologiya*, **19**, 193.
10. Andreyevsky, I. L. (1959). Geologichesky Sbornik 4, *Trudi VNIGRI*, **131**, 403 (Moscow: Gostoptechizdat).

11. Shturm, L. D. (1951). Pamyati Akad. I. M. Gubkina 275.
12. Senyukov, V. M., Yulbarisov, E. M., Taldykina, N. N. and Shishneva, E. P. (1970). Mikrobiologiya 39, 705.
13. Coty, V. F. (1975). The Role of Microorganisms in the Recovery of Oil. National Science Foundation, Easton, Maryland, 9–14 XI. 1975, 77.
14. Bond, D. C. (1961). US 2,975,835.
15. Jones, L. W. (1967). US 3,332,487.
16. Hitzman, D. O. (1959). US 2,907,389.
17. Hitzman, D. O. (1962). US 3,032,472.
18. Hitzman, D. O. (1965). US 3,185,216.
19. Hitzman, D. O. (1967). US 3,340,930.
20. Hitzman, D. O. (1972). US 3,650,326.
21. Wegner, E. H. and Stratton, C. A. (1971). US 3,598,181.
22. LaRiviere, J. W. M. (1955). Antonie van Leeuwenhoek J. Microbiol. Serol., 21, 1.
23. LaRiviere, J. W. M. (1955). Antonie van Leeuwenhoek J. Microbiol. Serol., 21, 9.
24. Bailey, N. J. L., Jobson, A. M. and Rogers, M. A. (1973). Chem. Geol., 11, 203.
25. Byrom, J. A. and Beastall, S. (1971). In: Microbiol., Proc. Conf., ed. P. Hepple, pp. 73–86. Institute of Petroleum, London.
26. Ekzertsev, V. A. (1958). Mikrobiologiya, 27, 626.
27. Foster, J. W., Cowan, R. M. and Maag, T. A. (1962). J. Bacteriol., 83, 330.
28. Heyer, J. and Schwartz, W. (1970). Z. Allgem. Mikrobiol., 8, 545.
29. Heyer, J. and Schwartz, W. (1970). Z. Allgem. Mikrobiol., 8, 565.
30. Jones, J. G. (1969). Arch. Microbiol., 67, 397.
31. Kalish, P. J., Stewart, J. A., Rogers, W. F. and Bennett, E. O. (1964). J. Petroleum Technol., 16, 805.
32. Knösel, D. and Schwartz, W. (1954). Arch. Mikrobiol., 20, 362.
33. Kolesnik, Z. A. and Shmonova, N. I. (1957). Doklady Akad. Nauk. SSSR, 115, 1197.
34. Krasilnikov, N. A. and Koronelli, T. V. (1974). Prikl. Biokhim. Mikrobiol., 10, 573.
35. Kuznetsov, S. I. (1967). Proc. VII World Petroleum Congr. Mexico. VII (2), 171. Elsevier, Amsterdam.
36. Lukens, H. B. and Foster, J. W. (1963). Z. Allgem. Mikrobiol., 3, 251.
37. Myers, G. E. and Samiroden, W. D. (1967). Producers' Monthly, 4, 22.

38. Norenkova, I. K., Brezhneva, I. V., Svechina, R. M., Karpenko, M. N. and Kulikova, E. M. (1976). *Mikrobiologiya*, **45**, 724.

39. Schwartz, W. (1972). *Naturwiss*, **59**, 356.

40. Simakova, T. L., Kolesnik, Z. A., Strigleva, N. V., Norenkova, I. K. and Shmonova, N. I. (1968). *Mikrobiologiya*, **37**, 233.

41. Smik, S. (1970). *Postepi Mikrobiologii* IX (1), 121.

42. Traxler, R. W. and Flammery, W. L. (1968). In: *Biodeterioration of Materials* ed. H. Walters, and J. Elphick, pp. 44–54. Elsevier, Amsterdam–London–New York.

43. Walhäusser, K. H. (1969). *Erdöl-Erdgas Zeit.*, **85**, 14.

44. Zajic, J. E. (1969). *Microbiol Biogeochemistry*. Academic Press, New York & London.

45. N. V. De Bataafsche (1958). Dutch 89,580.

46. Dostalek, M. and Spurny, M. (1957). *Ceskoslov. Mikrobiol.*, **2**, 300.

47. Dostalek, M. and Spurny, M. (1957). *Ceskoslov. Mikrobiol.*, **2**, 307.

48. Dostalek, M., Spurny, M. and Rosypalova, A. (1958). Prace ustavu pro naftovy vyzkum, Publ. 9, 29.

49. Dostalek, M. (1965). Abh. Deut. Akad. Wiss. Berlin, Kl. *Chem. Geol. Biol.*, 1965(2), 81 (Publ. 1966).

50. Dostalek, M. (1965). Abh. Deut. Akad. Wiss. Berlin, Kl. *Chem. Geol. Biol.*, 1965(2), 61 (Publ. 1966).

51. Kuznetsov, S. I., Ivanov, M. V. and Lyalikova, N. N. (1963). *Introduction to Geological Microbiology*. McGraw-Hill, New York.

52. Ekzertsev, V. A. (1960). *Geokhimiya*, p. 362.

53. Jaranyi, I. (1968). *M. All. Földtani Intezet Evi Jelentese Az 1968 Evröl*, 423.

54. Jaranyi, I., Kiss, L. and Szalanczy, G. (1967). *M. All. Földtani Intezet Evi Jelentese Az 1968 Evröl*, 345.

55. Dienes, M. and Jaranyi, I. (1973). *Köolaj es Földgaz*, **6**, 205.

56. Karaskiewicz, J. (1964). *Nafta (Katowice)*, **20**, 61.

57. Karaskiewicz, J. (1965). Abh. Deut. Akad. Wiss. Berlin, Kl. *Chem. Geol. Biol.*, 1965(2), 63 (Publ. 1966).

58. Karaskiewicz, J. (1966). IX International Congress for Microbiology, Moscow, 24–30 VII 1966. Abstract No. 261.

59. Karaskiewicz, J. (1968). *Nafta (Katowice)*, **24**, 198.

60. Karaskiewicz, J. (1975). *Nafta (Katowice)*, **31**, 144.

61. Karaskiewicz, J. (1977). *Isvestia Akad. Nauk. SSSR.*, ser. *Biol.*, **5**, 790.

62. Karaskiewicz, J. (1974) Prace Instytut Naftowego, 3. Katowice: Wydawnictwo *Slask*.

63. Jaranyi, I., Kiss, I., Szalanczy, G. and Szolnoki, I. (1963). Vorträge III Intern. Wiss. Konf. Geochem. Mikrobiol. *Erdölchem.* (*Budapest*), **2**, 633.
64. Jaranyi, I., Kiss, I. and Szalanczy, G. (1965). Abh. Deut. Akad. Wiss. Berlin, Kl. *Chem. Geol. Biol.*, 1965(2), 69 (Publ. 1966).
65. Hitzman, D. O. (1975). Statement made at the International Conference on the Role of Microorganisms in Oil Recovery, Easton, Maryland, (9–14 XI 1975).
66. Dunlop, D. D. (1976). In: *Microbial Energy Conversion*, a seminar sponsored by UNITAR and BMFT, Göttingen, 4–8 X 1976, 301. Erich Goltze KG, Göttingen.
67. Lazar, I. (1978). *European Symposium on Enhanced Oil Recovery*, ed. J. Brown, 5–7 VII 1978, p. 279. Heriot-Watt University, Edinburgh (publ. 1979).
68. Lazar, I. (1976). *Mine, Petrol, Gaze* (*Bucuresti, Romania*), **27** (10), 475.
69. Grigoriu, A. and Lazar, I. (1977). *Lucrarile primului simpozion de microbiologie industriala*, Iasi 18–19 XII 1976, 270. Universitatea 'Al. I. Cuza', Iasi, Romania.
70. Lazar, I., Balinschi-Zamfirescu, I. Dumitru, L., Grigoriu, A. and Mihoc, A. (1976). *Rev. Roum. Biol.*, *Ser. Bot.*, **21**, 53.
71. Mihoc, A., Popea, F. and Lazar, I. (1977). *Lucrarile primului simpozion de microbiologie industriala*, Iasi 18–19 XII 1976, 277. Universitatea 'Al. I. Cuza', Iasi, Romania.
72. Balinschi-Zamfirescu, I., Lazar, I., Dumitru, L., Mihoc, A., Grigoriu, A. and Popea, F. (1977). *Lucrarile primului simpozion de microbiologie industriala*. Iasi 18–19 XII 1976, 262. Universitatea 'Al. I. Cuza', Iasi, Romania.
73. Lazar, I. and Balinschi-Zamfirescu, I. (1977). Patent application to OSIM No. 91437, 20 August, 1977.
74. Lazar, I. (1978). Certificat de inovator nr. 44, Ministerul Educatiai si Invatamintului, Romania.
75. Lazar, I., Dumitru, L., Balinschi-Zamfirescu, I., Grigoriu, A. and Mihoc, A. (1977). *Lucrarile primului simpozion de microbiologie industriala*, Iasi 18–19 XII. 1976, 297. Universitatea 'Al. I. Cuza', Iasi, Romania.
76. Myers, G. E. and McCready, R. G. L. (1966). *Canad. J. Microbiol.*, **12**, 477.
77. Lazar, I. and Balinschi-Zamfirescu, I. (1977). *Lucrarile primului simpozion de microbiologie industriala*, Iasi 18–19 XII. 1976, 284. Universitatea 'Al. I. Cuza', Iasi, Romania.

78. Lada, A. (1959). *Oil and Gas Journal*, **57** (16), 93.
79. Lada, A. (1959). *Producers' Monthly*, **23** (6), 35.
80. Raleigh, J. T. and Flock, D. L. (1965). *J. Petroleum Technol.*, **17**, 201.
81. Allred, R. C., as reported by Davis, J. B. (1965). *Bacteriol. Revs.*, **20**, 261.
82. Amstutz, R. W. and Reynolds, L. C. (1963). *J. Petroleum Technol.*, **15**, 1073.
83. Karaskiewicz, J. (1968). *Biul. Inst. Naft.*, **18**, 1.
84. Sharpley, J. M. (1961). *Petroleum Engr.*, **33** (2), 55.

CHAPTER 5

The Exploitation of Deep and Offshore Fields

5.1. SPECIAL PROBLEMS OF DEEP AND OFFSHORE FIELDS

5.1.1. High Temperature

As the easily-exploited fields on the earth's surface become depleted we must turn to more difficult fields to maintain production. Often these will be unfavourable by virtue of their great depth and, since temperature increases with depth in the earth by roughly $0.014°C/m$, high temperatures will present an additional problem. As examples, in the North Sea representative temperatures might be 90 °C for the Forties field and as high as 120 °C for Magnus. For most forms of life these temperatures would be almost instantly fatal and we must consider carefully whether there is any possibility of using microbial enhancement techniques under such conditions, and if so what are the likely limits. Since even the deep wells have a considerable temperature range there is clearly no minimum useful temperature but the higher the upper limit, the greater the scope for the application of microbial techniques. Again it is not at all clear to what extent an initial reservoir temperature of, say, 90 °C will be maintained during waterflooding and some reservoir engineers have suggested that flooding with water at perhaps 5 °C might produce a significant and long-lasting reduction in reservoir temperature (L. P. Coombes and D. D. Dunlop, personal communications). It must be admitted that this view is not unanimous, but the outcome can be readily calculated given the appropriate data.

5.1.1.1.   Bacterial Growth at High Temperatures in the Laboratory
The general nature of bacterial response to temperature is well
known. Bacterial growth rates increase with increasing temperature up to an optimum while above this a further increase in
temperature leads to a rapid decrease in the rate of growth. As a
result, the maximum temperature which will permit growth may
be only a few degrees above the optimum. The minimum temperature capable of supporting growth will usually be well below the
optimum. The optimum temperature is not absolutely fixed for a
given bacterium but may vary slightly with the nature of the
growth substrate and with environmental factors such as pressure
and pH.

Optimum temperatures vary considerably from one organism to
another and bacteria are grouped into three main physiological
categories: psychrophiles, mesophiles and thermophiles, according to their optimum growth temperatures (Table 5.1).

TABLE 5.1
Temperature ranges for bacterial growth

Group	*Temperature °C*		
	Minimum	*Optimum*	*Maximum*
Thermophiles	40–45	55–85 (?)	60–100 (?)
Mesophiles	10–15	30–45	35–50
Psychrophiles	−5–+5	15–20	19–22

Clearly the thermophiles are likely to be of most interest for use
in enhanced recovery procedures, particularly those in deep wells,
and we must next consider whether even this group might be
sufficiently thermotolerant to meet our requirements. The upper
limits for both the optimum and maximum temperatures of thermophilic organisms are at present uncertain. A microbiologist will
accept that a bacterium is capable of growth at a certain maximum temperature if he can grow it at that temperature in the
laboratory in pure culture and can repeatedly subculture from one
growth vessel to another. It should be possible to demonstrate
growth by increase in cell members, increase in dry weight of cells,
increase in cell protein or by all three measurements. The highest

temperature for which these criteria have been met would appear to be 85 °C where reproducible growth of various strains of the genera *Bacillus*, *Sulfolobus*, *Thermomicrobium* and *Thermus* has been reported (references 1–4 and Atkinson, personal communication).

5.1.1.2. Growth at High Temperatures in Natural Environments

An upper growth limit of 85 °C, although useful, would still place considerable restraint on the exploitation of deep reservoirs. Fortunately, there is good reason to believe that bacterial growth is possible even above this temperature. The growth of living organisms in volcanic hot springs, especially those in Yellowstone National Park, USA, has been reported over many years. Brock and his colleagues have made a long and detailed study of the microbial life at the highest temperatures. Although they were unable to culture organisms at temperatures above 85 °C they obtained good evidence that bacteria could and did grow at significantly higher temperatures. For instance in a survey of almost 300 hot springs in the USA, New Zealand, Iceland and Japan they found that virtually every spring in the neutral and alkaline pH range had bacteria present, although bacteria tended to be absent from springs which contained the highest temperatures and low pH values. In this survey the water temperatures ranged up to 101 °C and five springs in the range 99–101 °C showed evidence of bacterial growth. In each case the presence or absence of bacteria was detected microscopically. When bacteria were present there were usually large numbers attached to solid surfaces. Very few were present in suspension in the water.[5]

   In this sort of study it is critical to establish that the bacteria present are really growing and are not non-growing contaminants from outside the thermal pool. Various lines of evidence suggest that the latter was not the case. The ability of bacteria to grow *in situ* was established by suspending microscope slides in the pools for periods from a few hours upwards. These quickly developed a layer of firmly attached bacteria. In another study growth rates were estimated from the rate of increase in cell numbers on slides suspended in two pools in Yellowstone Park, with temperatures ranging between 92·5 and 94·8 °C. Generation times of between 2 and 7 h were calculated, growth on the slides being distinguished from the continued attachment of new cells by treating control slides with ultraviolet irradiation at intervals to kill the cells already attached; a procedure which virtually prevented the increase in numbers.[6] In another series of experiments, Brock *et al.* studied the temperature range for growth of organisms from a spring with a temperature range from 90 to 93 °C.[7] Coverslips were immersed in the spring until they had become coated with bacteria, and were then placed in solutions of nutrients containing isotopically-labelled substrates and incubated at different temperatures. The uptake of radioactive substrate was used as a measure of the rate of metabolism. Radioactive glucose, lactate, acetate, bicarbonate, glutamate, leucine, phenylalanine and thymidine were all taken up by bacterial cells with a broad optimum between 80 and 90 °C and significant uptake took place at 93 °C. Uptake was shown to be due to bacterial metabolism because it was abolished by mercuric chloride, formaldehyde, azide and streptomycin.

The field experiments described above were performed by a well-respected group with considerable experience in working with thermophiles. Adequate precautions appear to have been taken to avoid an erroneous conclusion and it seems reasonable to accept their conclusion that bacteria can grow at temperatures up to about 100 °C. On the other hand members of the same group, and others, so far appear to have failed to grow bacteria above 85 °C in the laboratory. It is pertinent to ask why this discrepancy exists between the two sets of results since the ability to culture bacteria at the higher of the two limits would make a considerable difference to the range of wells which could be exploited by bacterial techniques. Tansey and Brock suggest that the difference may be due to the difficulties of setting up continuous culture equipment to work at around the boiling point of water,[4] but another, perhaps more plausible explanation comes from the work of Heinen.[2] He isolated two thermophiles from a pool at 86–87 °C but was only able to obtain growth in culture up to 80 °C and 84 °C respectively. The maximum growth temperature depended on the substrate available and was higher with pyruvate than with other substrates tested. Heinen pointed out that the substrates he used were not present in the pool in significant concentrations and suggested that, had the 'natural' growth substrate been available, a still higher maximum temperature in culture would have been demonstrated. A final point is the fact that culturing an unknown bacterial species is not always entirely straightforward and success may only follow after considerable adjustment of the growth conditions by trial and error. On the whole it seems unlikely that there is any fundamental reason why the growth of a bacterium at 100 °C in a natural environment should not be reproduced under laboratory conditions.

5.1.1.3.   Nature of Microbial Thermotolerance

Recent biochemical studies of various categories of macromolecule from extreme thermophiles have begun to provide an explanation of how the organisms are able to survive at such high temperatures. Their proteins show subtle differences in structure when compared with analogues of similar function from mesophiles, and these changes result in a higher degree of intramolecular interactions leading in turn to a greater thermal stability. An increase in thermostability may result from a change in a single amino acid though it is likely that most differences are greater than that. Ionic, hydrophobic and hydrogen bond interactions make proteins from thermophiles more stable both to chemical and heat denaturation. Stabilising consequences, both for macromolecules and for the protein-synthesising machinery, also follow from the production of certain polyamines characteristically produced by temperature-resistant organisms.

Membrane fluidity is another aspect of heat resistance. The lipids in thermophilic membranes contain higher proportions of saturated and straight-chain lipids than do those of mesophiles, but many bacteria can vary the composition of their membrane lipids over quite a wide range and do so in response to growth temperature. This has the effect of achieving a constant membrane viscosity.

Thermophile nucleic acid differences seem to affect mostly

tRNAs; the base composition of DNA appears not to differ consistently from that of mesophiles, which probably already has sufficient thermostability at suitable ionic strengths to account for its ability to function at the growth temperatures characteristic of thermophiles. Unique thermal stabilities are observed for tRNAs. In *Thermus* tRNA contains a high proportion of guanine and cytosine in the base-paired regions, which permits more hydrogen bonding and greater stability. These regions also contain more thiolated thymidine, providing a stronger stacking force within the molecule; indeed, the extent of thiolation increases with growth temperatures. A recent review discussing the whole question of extreme thermophily has been published by Zeikus.[8]

All of these differences are specific and are presumably encoded in the thermophile genotype. It seems likely that many genes in thermophiles would thus differ from those coding for similar functions in mesophiles. Nevertheless, it has been reported that DNA from a thermophile could transform a mesophile and confer on the recipient the ability to grow at higher temperatures (see section 5.2.4.4). Perhaps in this case growth in the mesophile was limited primarily by a single thermosensitive component, and by replacing this in the mesophile with an alternative version more resistant to high temperature, the recipient organism was able at one bound to improve its temperature tolerance until it encountered the next barrier. Whether repeated mutagenesis and selection for growth at ever increasing temperatures is a useful technique for removing such barriers successively, thereby converting mesophiles to thermophiles, has not been explored as far as we are aware.

5.1.2. High Pressure

Pressure increases in the earth with depth at an average rate of about 1 atm for every 3 m depth. The exploitation of deep wells using bacterial techniques thus has to contend with high pressures as well as high temperatures. As an example, well pressures can exceed 200 atm in the Forties field and 400 atm in the Magnus field of the North Sea. Bacterial cells injected in the waterflood may be subject to pumping pressures in excess of those due to the overburden, although these will be kept within limits in order to avoid inadvertent hydraulic fracture of the formation. There could be some cyclic fluctuation of the excess pressure due to the pumping action. This seems unlikely to cause a serious problem and in any case it could be considerably reduced if necessary by the use of multi-stage pumps.

5.1.2.1.   Pressure Limits for Growth

Bacteria vary considerably in their growth response to high pressure (for reviews see ZoBell and Kim;[9] Marquis;[10] Marquis and Matsumura[11]) and organisms are grouped into three broad categories on this basis. Relatively barosensitive organisms are unable to grow at pressures much above about 300 atm. The majority of organisms tested fall into this category and would therefore appear to be unsuitable for use in the deep wells (but see the comments below on the effects of other conditions on pressure stability). Relatively barotolerant strains will grow up to pressures of about 600 atm, while the relatively rare highly barotolerant strains will grow up to pressures of 1000 atm and above. The latter would clearly be capable of growth at even the most extreme reservoir pressures which we need consider at the present time. Whether any of them occurring naturally would combine all of the characteristics required for use in enhanced oil recovery seems rather doubtful. Methods by which other characteristics can be introduced into a barotolerant strain are discussed later.

5.1.2.2.   Physiological Changes due to Pressure

Even though an organism may be capable of growth, high pressure is likely to have a number of effects on its physiology, some of which could be detrimental to enhanced oil recovery.

Thus, organisms growing at elevated pressures may suffer reduced rates of growth and increased lag phases before growth starts.[10] A number of authors report effects of pressure on the yield of cells obtained from a given quantity of substrate. ZoBell and Budge compared the yield of cells of 22 relatively barophobic species and 8 barotolerant species of bacteria grown over a range of pressures.[12] In each case the yield of cells decreased steadily with increasing pressure. Yields at 600 atm were less than 50% of that at 1 atm in the case of the barotolerant species, and less than 10% in the case of the relatively barophobic species.

By contrast ZoBell reported that four of his bacterial strains gave higher growth yields at 100 or 200 atm than at 1 atm and Marquis and Keller found that *Escherichia coli* grew faster and produced more cells at 200 atm than at 1 atm.[13,14] Kriss and Mitskevich working with a marine pseudomonad found a higher growth yield at 350 atm than at 1 atm.[15]

Careful consideration must be given to the effects of pressure on growth rate and yield as they might result in delays in the bacterial release of oil or possibly even lowered bacterial activity to the point where the expected benefits are not obtained at all. Clearly though, there is at least a possibility that the effects of pressure might be beneficial in appropriate cases.

Several reports have described changes in bacterial morphology induced by high pressure. These include the induction of spore formation, variations in cell size, shape and particularly an interference with cell division in *Escherichia coli* resulting in growth as elongated filaments.[16-19] Changes of this sort, especially filament formation, could have a very serious effect on the ability of bacteria to penetrate through the reservoir rock and would greatly increase the chance of plugging. Fortunately, pressure induces morphological changes only in some bacteria.[10,19] Even different strains of the same species, *Escherichia coli*, reacted differently.[18,20]

It has already been mentioned (Chapter 3) that the ability of some bacteria to swim by lashing movements of their flagella may be significant in assisting intimate distribution throughout the reservoir rock. High pressure inhibits bacterial swimming.[13,21] The formation of new flagella is particularly sensitive to pressure and is completely inhibited by pressures of 200 atm in a number of bacterial species.[21] So far there appear to be no exceptions to this effect among the bacteria examined and no success has yet been obtained in attempting to isolate mutants which retain their motility at high pressure (Marquis, personal communication). Recently, Marquis and Bender have been able to achieve growth of *Streptococcus faecalis* at 1000 atm by serial subculture in agar stabs at pressure increments of 50 atm. A stable barotolerant mutant was characterised and was already present in the parental population.[22] Misshapen cells were observed, especially on the first subculture, but the morphology of the variant was normal.

5.1.3. High Salinity

Most offshore reservoirs will have no convenient access to large quantities of fresh water for use in pressure maintenance or flooding, and sea water or produced water will have to be used. Even on land produced water will frequently have to be reinjected: either there may be no other acceptable way of disposing of it or there may be a shortage of fresh water or the presence of clay in the reservoir rock might preclude the use of fresh water. Most bacteria commonly grown in culture or isolated from soil or fresh water are inhibited to a greater or lesser extent by the presence of concentrations of sodium chloride in excess of about 0·2 M (just over 10 000 ppm). Other solutes have a similar effect. Inhibition is due, at least in part, to the reduction in water activity caused by the dissolved solute (for a review see Kushner[23]). Organisms of this type are known as non-halophiles. Their growth would be prevented or severely inhibited by the high salt content of sea water (about 0·5 M) and of many reservoir brines.

Two other major categories of response to high salt concentration are recognised. Halotolerant organisms are capable of growth at high salt concentrations although growing best, or at least as well, at low concentrations. Halophiles are organisms which actually grow best at high salt concentrations. In some cases they have an absolute requirement for a particular concentration of ionic solute which can be fulfilled by a number of salts. The most extreme halophiles are remarkable organisms which can grow in saturated sodium chloride solutions (about 5·2 M, 320 000 ppm) and require a minimum sodium chloride concentration of about 2·5 M.[23-25] However, the extreme halophiles are not numerous and they have some peculiar physiological properties. They do not appear to be closely related to most other groups of organisms, which would make transfer of genetic information more difficult, and their absolute requirement for a high salt concentration might be an embarrassment. On the whole they do not seem the most promising source of isolates for use in enhanced oil recovery.

The halotolerant or moderately halophilic organisms include a larger number of organisms of fairly diverse relationships and many of them are easily capable of growth at the highest salt concentrations likely to be experienced in reservoir brines.[23,26] They lack the physiological peculiarities of the extreme halophiles and are often closely related to non-halophiles. Indeed, there has frequently been some controversy as to whether they are in any permanent way distinct from non-halophiles or whether they are merely adapted to grow in a saline environment. It now seems clear that, despite their close relationships with non-halophiles, the moderate halophiles and halotolerant strains are genetically distinct and have the sorts of properties most useful for our purpose.

5.1.4.   The Effects of Adverse Conditions in Combination

The effects of the adverse reservoir conditions discussed above cannot be considered entirely in isolation since they interact with one another and are also affected by pH and the nature of the available nutrients.

Thus ZoBell and Johnson studied the growth of more than 30 bacterial species at different temperatures and pressures.[27] None grew at 600 atm and 30 °C although four strains grew at this pressure when the temperature was raised to 40 °C. Marquis and Matsumura found that *Escherichia coli* would grow at a maximum pressure of 500 atm at 40 °C but only at 50–100 atm at 9 °C or 49 °C.[11] In general they observed that pressure tolerance was greatest at a temperature just above the optimum; at still higher temperatures pressure tolerance decreased markedly. The most remarkable result was that of ZoBell who succeeded in growing a culture at 105 °C and 1000 atm pressure.[28] Unfortunately attempts to subculture were unsuccessful.

Albright and Hanigman studying the growth of *Vibrio marinus* and other marine bacteria found that the maximum growth pressure increased with increased sodium chloride concentration while Novitsky and Kushner reported that the response of *Planococcus halophilus* to sodium chloride was changed on varying the growth temperature.[29,30]

Temperature tolerance is affected by the growth substrate. Thus Heinen growing two isolates from hot springs in culture found that their maximum growth temperatures were higher on pyruvate than on sucrose, glucose or acetate.[2] There is some evidence that temperature tolerance may be affected by pH. Brock and Darland in an extensive survey of bacteria in natural hot springs, found bacteria present in most of the springs sampled, including some with temperatures ranging up to 101 °C, but there was an absence of bacteria in those springs which combined the highest temperatures with low pH.[5]

These samples serve only to show that interaction does exist. The effects of these interactions on growth are likely to be complex and somewhat unpredictable. So far there are few signs of any general rules emerging and indeed responses differ considerably from one organism to another. There is no way of predicting how a given organism will behave in particular combinations of pressure, temperature, salinity, pH and nutrient availability. The only way to find out is by experiment.

5.1.5. What are the Environmental Limitations on the Use of Microorganisms?

We can see from what has been written above that there should be no problem in finding organisms which can tolerate even the most extreme pressures or salinities likely to be experienced under reservoir conditions. High temperatures obviously pose more of a problem. Strains able to grow at 85 °C are well known. An optimist would probably feel that operation up to a maximum of 95–100 °C might be possible, since bacterial growth is known to occur in this range in some natural environments. Probably few microbiologists would expect to be able to operate above 100 °C and this would clearly place definite and significant limitations on the reservoirs which could be exploited using microbial techniques.

Can resistance to high temperature, high pressure and high salinity be obtained in a single organism? No such organism seems to have been fully described but clearly the organisms which have been found in oil and brine produced from North Sea wells will be promising candidates for investigation.

5.1.6. Other Problems of Deep and Offshore Reservoirs

Due to the greatly increased costs of drilling, deep reservoirs are almost invariably associated with widely-spaced wells. The additional cost of building and operating platforms for offshore drilling and production is a further factor tending to increase well spacing. One consequence of this is that due to the sparsity of core samples, information about reservoir parameters is less detailed and there is a greater chance that significant changes in conditions will occur somewhere between the injection and production wells. Clearly, in order to operate a microbial enhanced recovery programme under these circumstances one would need to have great faith in the versatility and stability of the microbial system under a fairly wide range of conditions.

With the greater distances between injection and production wells will come greater delays in time and several years may elapse between the injection of water into the formation at one point and its appearance elsewhere. At present we know very little about the rate of movement of bacteria through permeable rocks but intuition and what evidence we have suggest that bacteria will move significantly more slowly than the waterflood itself. This might mean that the time delay in a microbial enhanced recovery technique would be longer than in a corresponding chemical flooding operation and that the delays might be such as to make the whole operation uneconomical even if appreciable enhancement of recovery were eventually obtained. We cannot yet evaluate this effect. It may even be the case that bacteria will move much faster than we expect or that the products of bacterial metabolism move ahead of the cells, carried by the waterflood, so that no appreciable extra delay occurs.

Long delays in time will also make great demands on the stability of the system. The properties of the bacterial culture could certainly be very precisely controlled up to the point at which it was injected, but after that no control would be possible. The system could break down in at least two ways. For instance nutrients injected with the waterflood could be utilised by competing bacteria inadvertently introduced at the same time. Although control measures are taken to minimise the introduction of unwanted bacteria into the formation it is clearly not practicable to achieve complete success under field conditions. Another possibility is that the bacteria may undergo mutation and lose some property essential to the success of the operation. The most likely property to be lost in this way would be the production of a surfactant or polymer. Such a substance would probably not be essential to the continued growth of the organism. Therefore nonproducing mutants would still be able to grow and might actually be at a competitive advantage by not carrying the burden of producing it. 'Mutation is a continuous process in any bacterial culture, although normally proceeding at a very slow rate. The effective loss of a small fraction of the cells would not be significant. However there is a distinct possibility that under reservoir conditions mutation rates might be unusually high. Atkinson (personal communications) detected greatly increased mutation frequencies in some loci when *Bacillus stearothermophilus* was grown at the higher end of its growth range. Increased mutation frequencies coupled with selective pressure in favour of the mutant could lead to the breakdown of the system before any appreciable benefit had been obtained.

Fortunately, the genetic stability of a bacterial system can be tested fairly adequately under laboratory conditions so the general likelihood of genetic breakdown during the very long time intervals can be estimated with some confidence. Clearly the system must be designed with great care to have a high degree of genetic stability under the circumstances to be encountered in the reservoir, and in order to achieve this aim a very high degree of microbiological, genetical and biochemical insight will be essential.

5.2. HOW TO OBTAIN A SUITABLE BACTERIAL CULTURE FOR ENHANCED OIL RECOVERY

5.2.1. Shortcomings of Previous Approaches

In the past two types of approach have been used to the problem of finding a microbial culture for use in enhanced oil recovery. One has been to use pure cultures of organisms already available or isolated for the purpose.[31-34] The second approach used widely in the Eastern European trials has been to select cultures for the purpose by empirical procedures. The resulting cultures have consisted not of single organisms but of groups of organisms, or consortia, operating together to achieve the desired end result (Chapter 4). Despite the fact that both types of procedure have achieved a limited degree of success we feel that an entirely different approach is now called for. It is most unlikely that any single strain could be found in nature which combines together an ideal set of characteristics for mobilising crude oil in reservoirs. Further, in selecting a functioning consortium by the empirical techniques used by the Eastern European workers (Chapter 4), it is not likely to be possible to select simultaneously for all the desirable characteristics.

The approach we favour is a more fundamental one and consists of two stages. The first stage is to identify an ideal set of characteristics for a bacterium to be used to enhance oil recovery. The second stage is to consider how these desirable characteristics could be obtained in bacterial cultures, and how they could be combined together. It is, of course, most improbable that any single microbial strain or consortium will ever be suitable for enhanced oil recovery in all possible situations. More probably the enormous variation between fields in factors such as oil composition, oil saturation, rock structure and chemistry, temperature and pressure, chemical composition of the reservoir brine and waterflood, taken together with deliberately injected substances (such as biocides and corrosion inhibitors), will require the tailoring of a series of microbial strains. However, we can usefully consider the general nature of the properties needed by an ideal bacterial strain and proceed beyond that to a broad discussion of ways of acquiring the desired strain or mixed population of strains.

5.2.2.   The Ideal Organism

5.2.2.1.   Nutrition

In the field trials conducted so far molasses has been injected into the formation in nearly every case. Molasses is a comparatively cheap, readily available material which, because of its high sugar content is utilised rapidly by a wide range of organisms under either aerobic or anaerobic conditions. Under anaerobic conditions, which seem certain to predominate in reservoirs during microbial enhanced recovery techniques, sugars would give rise to a range of products some of which (e.g. methane, carbon dioxide, hydrogen and various organic acids) are of potential usefulness in promoting the mobilisation of oil.

It is not always entirely clear exactly what role has been played by the molasses used in field trials. In the trial conducted by Coty, Updegraff and Yarbrough (Yarbrough, personal communication; see also ref. 34) molasses was injected continuously with the waterflood at a concentration of 2%. In this case the metabolic products from molasses predominated (e.g. 80% of the carbon dioxide sampled from the production well was derived from the molasses). On the other hand some of the Eastern European trials used comparatively low quantities of molasses.[35,36] In these cases it is uncertain whether sufficient metabolic products could have been derived from the molasses materially to affect oil release, although instances were recorded (see Chapter 4) in which a decline in the enhanced rate of oil recovery following the injection of bacteria was restored to its earlier level by the injection of a second charge of molasses. This does suggest an important role for molasses in those trials. Furthermore, the wells employed so far have always had a low residual oil saturation and it seems quite likely that the quantity of metabolic products required may be related to the quantity of oil produced.

In any case it seems out of the question that it would be economically feasible to use molasses as the direct source of the

major bacterial metabolic products in enhanced recovery prog-
rammes with deep wells, or those some distance offshore, because
of the enormous quantities required and the long delay times
before any return on the investment could be expected. For this
type of situation the requirement for added nutrients must be
minimised. The obvious way to achieve this is by using one or
more components of the crude oil as the source of carbon and
energy for bacterial growth and metabolism.

It is well known that many bacteria will break down a variety of
crude oil components in the presence of oxygen and utilise them
as a source of carbon and energy.[37-40] Unfortunately, aerobic
metabolism will not be possible in the reservoir due to a severe
limitation in the supply of oxygen. The solubility of oxygen in
water is low and oxygen in the waterflood would presumably be
removed by bacteria growing near to the injection site. Further
the presence of oxygen would probably encourage the growth of
unwanted bacteria in that part of the formation near the injection
well. These would remove nutrients and might have other undesirable effects such as plugging of the formation (see Chapter 3).
Although the ability of many bacteria to metabolise crude oil
components under aerobic conditions is well established the
corresponding anaerobic process is much more controversial. The
few scattered reports claiming anaerobic metabolism frequently
lack detail and confirmation is needed.[41-43]

So far the reports of anaerobic oil metabolism do little more
than whet the appetite. The potential benefits of employing such a
system however, make it well worth investigating the possibilities
further. An interesting observation which might provide a starting
point is the report of Swain et al. that a strain of Pseudomonas
aeruginosa is capable of using various n-alkanes as sole carbon
source in the presence of oxygen, or of growing anaerobically at
the expense of carbohydrates with nitrate as electron acceptor.[39]
There seems no immediate reason in principle why the organism
could not be mutated to produce anaerobic growth on alkanes,
using nitrate as the electron acceptor.

It would not be necessary for the organism employed for oil
mobilisation to utilise all the components in crude oil; indeed, it

would be better in many ways if it did not. As a result the effects of metabolism on the properties of the crude, especially on viscosity, would be more predictable and the extent of growth would be regulated automatically. There would be no possibility of massive growth in one part of the reservoir using all the oil in that region, and perhaps interfering with the success of the operation either by removing nutrients to an excessive degree or by inducing blockage.

Apart from its obvious economic advantage, a system based on the anaerobic utilisation of crude oil is less likely to stimulate the growth of unwanted bacteria in the reservoir than one based on the injection of a substrate which is usable by a wide range of microorganisms. The ability to grow anaerobically at the expense of crude oil without organic supplements is therefore the first characteristic of the ideal organism.

5.2.2.2. Mobilisation of Oil

While growing on crude oil, or some component thereof, as a carbon and energy source, the ideal strain must be capable of assisting the mobilisation of the crude oil. Some of the ways in which this might be achieved have been reviewed in Chapter 1. Most of them depend on the effects of the products of microbial metabolism. A considerable variety of products has been produced from oil components by bacteria under aerobic conditions (reviewed by Abbott and Gledhill) but we can only speculate on the products which may be produced under anaerobic conditions.[44] Methane and perhaps carbon dioxide are likely products which could be useful either for pressure maintenance or for the displacement of oil from blind capillaries or, especially in the case of carbon dioxide, for reducing oil viscosity and increasing its volume. Other potentially valuable products would be surfactants or polymers to increase the viscosity of the aqueous phase.[45,46] Studies on surfactant flooding have emphasised the specific properties required of a suitable surfactant (summarised by Stewart).[47] Surfactants produced in situ by bacterial metabolism will probably need to have a rather different set of properties for maximum effectiveness. The concentration of the product generated will also be important. Substances such as surfactants and polymers may well be produced in rather low yields by microorganisms. Possible methods for increasing the yields produced are discussed in section 5.2.4.1.

A number of microbial surfactants have already been characterised.[48] They are produced during growth on a variety of media and interestingly enough biosurfactant production is always found during growth on hydrocarbons. The organisms, of course, live in the aqueous phase and probably attack the hydrocarbon only at the water/oil interface. Presumably, by reducing the interfacial tension, the surfactants promote a degree of emulsification and thereby permit enzymic attack. These seem to be exactly the types of compound to be borne in mind in developing microbial systems with the specific goal of employing them for *in situ* oil mobilisation.

An alternative approach might be to attempt to modify the oil directly by hydrogenation or other form of reaction to increase its mobility.[49] This is likely to be a very slow process since any bacterial action could take place only at the periphery of the oil phase where it is in contact with water. The volume of the average oil droplet and the area of available surface for bacterial action would be important factors in determining reaction rates.

5.2.2.3. *Mobility in the Reservoir*

The ideal organism must be capable of penetration throughout at least the major part of the reservoir (see Chapter 3). The minimum extent of penetration needed may vary according to the mechanism by which oil is mobilised. For example, surfactant generated in one location may mobilise oil in another part of the reservoir, but if oil mobilisation were dependent upon the selective degradation of viscous components this could only be achieved in locations actually reached by the bacteria.

The factors which would govern the mobility of bacteria through porous reservoir rock are not precisely known. A comparatively small size and a nearly spherical shape will probably be important and there should be no tendency for cells to form clumps or chains or to produce extracellular sheaths or slimes which would hinder motility. The ability which some bacteria possess to swim independently by means of flagella could also be important since this might counteract any tendency for cells to settle out from slowly moving currents of water and might permit penetration to rock pores or capillaries not swept by the water-flood. Many microorganisms have the ability to detect concentration gradients of nutrients and other substances and swim towards their source. This well-known phenomenon is called chemotaxis. If microorganisms exist which can seek out crude oil components in this way, the property could be very useful in mobilising oil in localised regions of the formation.

Hitzman has suggested that spores might provide a better inoculum then vegetative cells because of their smaller size, absence of extracellular slimes and capacity to survive for extended periods in the absence of nutrients. The ability to form spores might be advantageous but since spores are non-motile, the use of a spore inoculum would sacrifice, for a time at least, the advantage of independent bacterial movement.[33]

5.2.2.4.   Undesirable Properties

There are certain undesirable properties which the ideal organism would lack. It would not cause plugging either by excessive growth in a localised region or by the production of slimes. Many bacteria autolyse after their death: that is, enzymes which have been produced by the cell break down the major cellular structures. Such a property might be valuable as a safeguard against plugging and to ensure that after the death of a cell many of the nutrients it contains would be released and be available for reuse.

The production of chemical precipitates such as sulphides, carbonates or ferric hydroxide would be undesirable because of the danger of plugging, while sulphide ion is particularly undesirable due to its corrosive properties.

5.2.2.5.   Tolerance of Reservoir Conditions

Deep wells are likely to have high temperatures, high pressures and probably high salinities; the organism of choice thus needs to be able to grow in any combination of these conditions and at any pH likely to be encountered in the chosen reservoirs. Since it would be inconvenient and expensive to grow the culture and manipulate it prior to injection under pressures characteristic of the reservoirs, and since the thermophilic culture must be injected together with relatively cold flood water, the ability to grow at atmospheric pressure and to survive for short periods at temperatures as low as 5 °C would be extremely useful for the most efficient operation.

5.2.2.6. Tolerance of Substances Injected with the Waterflood

The pretreatment of the flood water may include the addition of flocculating agents, biocides, oxygen scavengers and corrosion inhibitors. If reservoir brine is being reinjected it may also contain defoaming agents and emulsion breakers. The ideal organism would have to be able to survive and grow in the presence of the specific agents chosen.

The prevalent use of biocides presents special problems. An effective microbiologically-based enhancement method must ensure that the organism of choice, injected with the waterflood, is the one which forms the dominant microbial population even if some other organism should be present. It would nevertheless be useful if the numbers of potentially competing organisms, however disadvantaged they might be, were kept to a minimum. This might best be achieved by the employment of biocides as in current technology. If the chosen organism could be made resistant to the biocide, the latter could be introduced into the waterflood in the normal way; if not it would be necessary to employ a means of removing or inactivating the biocide after it had destroyed undesirable contaminants but before inoculating the waterflood with the desired organism. In the case of some biocides, such as the quaternary ammonium compounds, the chosen organism could probably be made resistant by mutation and selection because resistant bacteria are known to arise in nature. In the case of others, such as chlorine, bacterial resistance is not known to arise.

5.2.2.7. Genetic Stability

The time scale in the exploitation of deep wells and those situated offshore is likely to be comparatively long because of the wide well spacings, so a culture injected with the waterflood would need to retain its initial properties for many years in order to be useful. Although it is quite likely that the method would require the continued injection of fresh inoculum, the time delay before this reached remote parts of the field might be very long and there would be no way in which a culture which had degenerated could be replaced quickly. Cells which had degenerated by mutation could well be harmful rather than simply ineffective, since by depleting the supply of supplementary nutrients such as mineral salts or nitrogen, they could prevent growth of undegenerated cells. A high degree of genetic stability would therefore be an essential property.

5.2.3. Methods of Obtaining the Ideal Organism

5.2.3.1. Pre-existing Cultures

Very large numbers of microorganisms are maintained in culture by individual investigators and by culture collections whose main function is the maintenance and distribution of organisms on demand. There are many of these throughout the world. Examples are the American Collection of Type Cultures and the (British) National Collection of Industrial Bacteria. A guide to world culture collections has been produced.[50] Thus one way of obtaining an organism with a well-defined property is to search through the catalogues or write to the curators of appropriate culture collections. Another is to search the scientific literature for an organism with the desired properties and to obtain a culture from the individual researcher. Of course an organism able to grow under reservoir conditions and increase the mobilisation of oil would be of enormous commercial importance. It is unlikely that it would be available for the asking, and its very existence might well be kept secret, or it might be protected by patent.

5.2.3.2. Isolation from Nature

If no strain has been described with the desired properties it is frequently possible to isolate something suitable from a natural source. This would most probably be done by a technique known as enrichment culture, a process performed by setting up an environment which will favour the growth only of an organism with the desired properties and adding a natural material such as soil or water as a possible source of the organism required. If suitable conditions have been provided even a single cell with the desired properties will multiply and eventually dominate the culture. A series of subcultures is then performed to eliminate organisms which are unable to grow. Frequently, several different strains with the desired properties will be selected. For example, in order to obtain an organism able to grow at 95 °C a growth medium would be set up at that temperature and inoculated with a likely source of thermophilic organisms, such as water or rock from a natural hot spring at 95 °C or above. A more mundane source of inoculum might be water from domestic or laundry hot water systems.[52] Only strains with the desired characteristics would be able to grow and success would depend on the skill of the operator in choosing appropriate medium and culture conditions to select the bacteria sought from among the many other forms which would inevitably be present at the start of the experiment.

Organisms able to metabolise oil components under anaerobic conditions might be isolated by incubating a medium containing crude oil (or perhaps selected pure components of crude oil), together with a suitable electron acceptor and other mineral components. This would be inoculated with soil, mud or water samples as potential sources of the desired organism. Only organisms capable of growing at the expense of oil components would be present in large numbers after a series of subcultures. This technique could be extended to look for organisms capable of growth at the expense of crude oil, which could simultaneously mobilise oil from rock. In this case the crude oil might be supplied by immersing pieces of oil-saturated rock in the solution containing the other medium components. Success would be indicated in the first instance by the release of oil from the rock samples.

Barotolerant organisms could obviously be expected in samples of water and mud from high-pressure environments such as the deep sea where barotolerance would be essential for growth. On the other hand Kriss has reported that highly barotolerant bacteria can be isolated from samples of garden soil.[53] Kriss actually claimed that the bacteria in his garden soil grew better under pressure than did those from samples of deep-sea muds, and one strain was isolated which could grow at 850 atm. Since barotolerance is not limited to situations where it is a prerequisite for growth there would appear to be considerable scope for the isolation of barotolerant strains from a wide range of natural environments. Kriss and Zaichkin actually investigated the distribution of barotolerance in various types of soil.[54] Recent reports that bacteria are present in many of the crudes produced from North Sea wells suggest that these would be well worth investigating as sources for the isolation of bacteria tolerant of reservoir conditions. The chances of obtaining bacteria which are also able to metabolise crude oil components would also seem to be good.

Variations of the enrichment culture technique could be used to look for organisms having many of the individual characteristics required for enhanced oil recovery and some organisms with combinations of desirable characteristics could probably be isolated in this way. Isolation by enrichment culture is sometimes easy but success is by no means automatic. Given that an organism with the required characteristic really does exist somewhere in nature, failure could result from a number of causes, e.g. the absence of any suitable organisms in the particular inoculum employed, choice of conditions which did not sufficiently favour the desired organism relative to others, and unsuitable growth medium or conditions. Success would depend on the skill and experience of the operator, on the amount of time and effort expended, and to some extent, as with most scientific endeavours, on a measure of good luck.

For our purposes there are two serious limitations to enrichment culture. One is that the organism with a complete set of the desired characteristics may well not exist in nature at all, or, if it does exist, it may be very rare and not be present in any of the inoculum samples used. The other is that enrichment isolation will not be possible for all characteristics in which we are interested, especially the negative ones such as the non-production of slimes or hydrogen sulphide.

5.2.4. How Can the Desired Characteristics be Brought Together?

Our knowledge of microbiology strongly suggests that each of the characteristics needed for our ideal organism already exists in some microbial form or another, but the chance that all of them are now present within one single bacterial strain seems very small. Fortunately, there are well-studied techniques which may permit us to combine almost any set of characteristics which are biologically possible and not mutually exclusive.

5.2.4.1. Mutation and Selection

Any microorganism is subject to a continuous process of change by mutation. The frequency of this change in a given characteristic is likely to be low (about 1 in 10^6 cells) but since a dense bacterial culture could easily contain 10^9 cells/ml some mutants are invariably present. The natural frequency of mutation can be increased by a variety of chemical and physical mutagens such as treatment with certain derivatives of nitrosoguanidine or irradiation with UV light.[55] It is often possible to select and culture a single mutant with the desired characteristics from among large numbers of unwanted cells and the new characteristics will be a permanent feature of the mutant strain (see Chapter 2). One way in which this selection can be achieved is by spreading portions of a diluted cell suspension over the surface of solidified growth medium contained in petri dishes (plates). The dilution is arranged so that each cell is well separated from its neighbours. Each cell divides repeatedly giving rise to a colony of millions of genetically-identical individuals which becomes visible to the naked eye. Cells from a colony with desired characteristics can then be removed and cultured.

A bacterial strain might be unsuitable for enhanced oil recovery because it produced extracellular slime. In this instance mutant colonies which failed to produce slime could be detected by eye since the slime-producing colonies would probably appear smooth and glistening but the non-slime-producing mutants would have a rougher appearance.

Many desirable types of mutant would not be morphologically distinguishable. Geneticists have devised a variety of ingenious tricks for detecting these on agar plates (see, for example, references 55 and 56). Using such techniques mutants which fail to produce unwanted metabolic products such as hydrogen sulphide can be detected and isolated. By applying mutation and selection techniques in sequence several deleterious characteristics could be eliminated from a strain.

An important point, particularly relevant in the removal of certain properties from a bacterial strain, concerns the nature of the mutation. Mutations themselves are of two main types: substitution of one base at a particular position in DNA by another, and addition or removal of one or more of the bases. The first type of (substitution) mutation changes the nature of the information encoded by that section of DNA, and this may be expressed both in the genotype and the phenotype of the cell carrying the change, i.e. the cell is in some way different from its parents and the difference is inherited by its offspring. However, there is a degree of probability that the change will revert to its original state and this will occur with some frequency dependent upon the local conditions, nature of the mutation, etc. This means that a property lost in this way may be spontaneously regained at a time outside the operator's control and very much against his interests.

The second type of mutation results either in a change of the readout frame for information transfer from DNA, or the actual loss of a greater or lesser amount of information from a particular gene by excision. This last type of deletion mutation is by far the most satisfactory if a property is to be removed from an organism. When a large amount (perhaps many nucleotides long) of the information required for the expression of a property is actually physically lost from the chromosome the chance of its spontaneous reversion to the *status quo ante* becomes vanishingly small. There are mutagenic techniques in use which preferentially cause deletions and their presence can often be deduced from genetic analysis. Deletion mutations, because of their great stability, are to be preferred for our purpose in which it would not be possible to correct a biological defect in the bacteria once the latter had been injected into an oil-bearing stratum.

It is frequently an easy matter to select mutants resistant to particular chemicals in the growth medium and it should not be difficult to obtain mutants resistant to some of the biocides and other substances such as corrosion inhibitors which might be added to the waterflood. In this way specific bacteria could be introduced into the reservoir for enhanced recovery techniques without forgoing the normal precautions against the growth of unwanted organisms.

An interesting possibility is that mutation and selection might be used to enhance the tolerance of a bacterial strain to reservoir conditions such as high temperature, pressure or salinity. It must be recognised, however, that the scope for such change is at present uncertain. The possibilities of success depend upon the amount of genetic information which has to be altered in order to make a bacterium more thermotolerant, more barotolerant or more halotolerant. At present we do not know enough about the ways in which some bacteria are able to tolerate extreme conditions to know whether the amount of information required is great or small (but see section 5.1.1.3.). On the whole it would seem likely that very considerable changes in a cell would be necessary to increase its tolerance significantly. In this case one would not expect to achieve much success by mutation and selection in the limited time actually likely to be devoted to such an experiment (but see below).

So far attempts to alter pressure tolerance and salt tolerance of bacteria have yielded mainly negative results (see references 23–25, 57 and ZoBell, personal communication). However, it has been possible to increase the salt tolerance of a species of green alga by serial subculture so it would seem at least plausible that the same thing could be achieved with bacteria and there is indeed one report of a streptomycin-resistant bacterial mutant having increased barotolerance.[58,59]

Earlier attempts to increase the thermotolerance of bacteria were also generally unsuccessful. In one case a strain of *Bacillus subtilis* had its maximum growth temperature increased from 50 °C to 72 °C by increasing the growth temperature in steps of one or two degrees.[60] After growing the strain again at 50 °C the ability to grow at the higher temperature was lost. This then was not a case of genetic change but a matter of obtaining the maximum physiological adjustment of which the strain was capable. That, too, could be a valuable property. We are not aware of any experiments in which stepwise increase in growth temperature has been combined with repeated mutagenesis and selection.

5.2.4.2. Selection in Continuous Culture

Nowadays, microbiologists make extensive use of continuous culture of microorganisms. In the simplest type of system, the chemostat, growth medium is continuously pumped into the culture vessel at a constant slow rate. The vessel is stirred to produce homogeneous conditions, and a mixture of spent medium and cells is continuously removed at the same rate at which fresh medium is supplied. In this way the same culture can be kept growing without interruption for periods of weeks or months. Under appropriate conditions a steady state is set up in which growth and metabolism proceed but the chemical composition of the medium (levels of substrates and products) and the concentration of cells remain constant. Accounts of the theory of continuous culture have been presented by Tempest, Pirt and others.[61,62] A useful feature of continuous culture is that growth rate is controlled by the concentration of a single limiting nutrient in the fermenter. A mutant which is able to utilise the limiting nutrient more rapidly will thus have a selective advantage and will steadily increase in numbers until it finally completely displaces the wild-type. Further mutants which increase the efficiency of the attack on the limiting nutrient still more will also accumulate. The method can thus be used to select not for a single mutational step but for a series of mutants all having the same kind of effect in a cumulative process. A potential use of such a system would be to increase the efficiency of attack of a chosen strain on a particular crude oil component or perhaps to modify the substrate specificity of an enzyme in order to change the pattern of attack on crude oil. Clarke has written a useful review on the possibilities of changing enzyme levels, substrate specificities and control mechanisms by mutation and selection in batch and continuous culture.[63]

Continuous culture systems are also of great potential use for selecting strains resistant to adverse conditions such as inhibitory substrates. The advantage of continuous culture for this purpose is that the intensity of the adverse condition can be increased very slowly so that mutants slightly more resistant than the average cell are favoured and outgrow the original population. As the intensity of the adverse condition increases sequences of mutations build up in the population which thus becomes steadily more resistant.

5.2.4.3. Improved Techniques for Selection in Continuous Culture

A refinement of the technique is to monitor the cell density in the culture vessel by an optical device and to use the resulting signal, via a control system, to regulate the intensity of the adverse factor. Recently, some success has been obtained with a system of this sort in which increased temperature was the adverse factor. By raising temperature in very small increments until growth was adversely affected, a strong selection pressure was applied in favour of any mutations which increased thermotolerance. When such mutations had begun to predominate and growth rate again increased, the control mechanism instigated further small temperature increases until growth was once more inhibited.

In this way it has proved possible to extend the growth range of some species appreciably in a relatively short time. So far as we are aware no attempt has yet been made to use this system to extend the upper growth range of the extreme thermophiles. The practical problems of running a continuous culture system at temperatures approaching the boiling point of water are not entirely trivial but the results of such an experiment might prove of great significance in relation to the use of microorganisms in the deep wells.

It should also be possible to apply similar principles to the selection of halotolerant mutants. In this case the temperature would be constant and the control system would be used to regulate the salt content of the growth medium being pumped into the culture vessel. Since the difficulties of isolating mutant strains with significantly increased salt tolerance are probably due to the large amount of genetic information which needs to be changed rather than to any fundamental obstacle, an approach of this sort in which continuous selection pressure can be applied would seem to be well worth trying. Again it seems likely that barotolerant mutants could in principle be selected by such a technique, but the technical problems of running a continuous culture system at varying pressures of some hundreds of atmospheres for periods of weeks or months seem somewhat formidable although not perhaps insoluble.

In any case it seems that a programme of mutation and selection could well have a part to play in obtaining a suitable bacterial strain for enhanced oil recovery but it must be emphasised that there are practical limitations on what can be achieved. Selection, with or without prior mutagenesis, is based on genetic changes which are essentially random. The sort of changes which will occur involve loss or gain of one or more DNA base pairs or substitution of one base for another. Changes of this sort in the DNA can result in loss of function of an enzyme, changes in enzyme reaction velocity, alteration of substrate specificity or the affinity of the enzyme for its substrate, and alterations in control mechanisms of enzyme synthesis resulting in greater or smaller amounts of enzyme being produced. In turn these changes can result in a wide range of chemical, morphological and physiological changes in the cell. However, there are some changes which are not likely to result, such as the appearance of completely new enzyme activities. Such new activities would require the acquisition of extensive and highly specific new sequences in the DNA, changes which could come about only as the result of information transfer from another source. We could not therefore expect to use mutation and selection to endow a bacterium with the ability to produce a surfactant or degrade a specific crude oil component unless the organism already possessed the potential for these activities or for something very similar. On the other hand we would expect to be able to increase the yield of a surfactant or to improve the efficiency of attack on a substrate. A great deal of experience has been gained on methods of increasing product yields in microorganisms in applications such as antibiotic production. Based on this we would confidently expect to be able to increase the yield of a product such as a surfactant by quite substantial amounts. The limiting factors would be the time and effort available.

5.2.4.4. Information Transfer

If it proves impossible to bring about the changes required by mutation and selection, techniques may be developed to transfer characteristics from one organism to another by employing one of three classical tools of bacterial genetics: conjugation, transduction and transformation. These were discussed briefly in Chapter 2. A more detailed treatment is given by Hayes.[56] These processes all suffer from the limitation that they normally only work well to transfer information between closely related strains. Even so, they might well have a part to play in constructing the ideal organism. Thus, both Lindsay and Creaser and Friedman and Mojica-a have reported that a mesophilic strain of *Bacillus subtilis* could be transformed by DNA from a thermophile, *Bacillus caldolyticus*, to grow at temperatures above 70 °C.[51,64] This rather surprising observation suggests that it might be possible, given a strain with useful properties but a modest maximum growth temperature, to raise the temperature limit by transformation with DNA from an extreme thermophile. Perhaps, too, DNA from extreme barophiles and moderate halophiles can be used to extend the growth limits of related strains.

5.2.5. Combining the Desired Characteristics by Genetic Engineering Techniques

The techniques of genetic engineering discussed briefly in Chapter 2 and at more length by Atherton *et al.*, although perhaps more difficult to apply, offer in the long run far more flexibility and scope in tailoring an organism precisely to the desired characteristics.[65]

The strategy we propose would be to use one organism as a

genetic host. This should combine together as many of the desired properties as possible and should also possess a suitable plasmid into which DNA from other organisms could be incorporated. Properties such as the ability to grow at high temperature, pressure and salinity may well involve considerable amounts of genetic information. It might therefore be advantageous to use a host which already had these properties and to transfer other properties into it. These might include, for example, the ability to degrade oil components anaerobically and to produce surfactant, since these properties probably involve a comparatively small number of specific enzymes and therefore require a correspondingly small amount of genetic information to be transferred into the host organism.

We do not at present know of any organism which would be entirely suitable to act as the primary host. Many thermophilic organisms possess plasmids (Atkinson, personal communication) and some are known to tolerate fairly high concentrations of sodium chloride, but there seems to be no information at present on their response to high pressure. It is also uncertain whether enzymes from a mesophilic organism could operate successfully in a thermophile. A possible scenario is thus to use as a host organism one which is naturally able to cope with the environmental conditions of the reservoir, and to endow it genetically with such other properties of metabolite synthesis, etc., as may be desired. Those additional properites would originate from other bacteria, perhaps from mesophiles, since that group is probably the most widespread and certainly the best documented. Mesophilic properties may not survive at high temperatures when expressed in the thermophile recipient because they would not have evolved in nature to resist its denaturing effects. It seems likely, however, that by mutation and selection one could improve the thermotolerance of individual species of mesophilic proteins more readily than one could improve the whole mesophilic cell, using either selection pressure in a chemostat or sequential cycles of mutation and selection. In this way it might be possible to build up an organism capable not only of survival at high temperature, but able also to express at such temperatures properties originally derived from species able to exist only in much cooler circumstances.

There are clearly many problems to be solved before genetic engineering techniques can be applied to the construction of a microorganism for use in enhanced recovery processes but the general strategy to be used is fairly clear. A prerequisite for genetic engineering is a good knowledge of the properties to be transferred into the host. By the time we have this information we may also have suitable host organisms available and, judging by the present rate at which the techniques are being developed, the whole operation will have become very much easier to perform.

5.2.6. Combining the Desired Characteristics by the Use of Consortia

So far our discussion has assumed that the best way to set up a microbial enhanced recovery system would be to obtain a single organism with all the required characteristics. There is an alternative approach. Microbiologists frequently find that although complex organic compounds can be degraded at high rates in natural situations it can be very difficult to isolate bacteria which can reproduce the same effect in pure culture. The pure cultures which are isolated may be unable to carry out the complete degradation or they may be able to do so but only at very low rates compared with the natural population. In some cases at least the reason for this is that in nature degradation is carried out not by a single organism but by a community of different species. Microbiologists refer to such a microbial community as a consortium.

The interrelationships in a consortium are likely to be complex and there is probably no example in which they are all fully understood. Slater and Somerville describe a number of examples, in each case considering the interactions between the component species.[66] Despite their complexity, consortia can be stable over long periods of time. Thus, Ferry and Wolfe grew a consortium in the laboratory for three years on benzoate, while the anaerobic sludge digesters used in sewage purification, which depend on a complex consortium, can be operated with confidence for months or years.[67] This stability may seem surprising, but it is probably due to the fact that each organism in the consortium is dependent on others in some way and that any excessive growth of one member is self-limited leading to a stable equilibrium.

Some clue as to the types of interaction which might be significant can be obtained from those examples already analysed. Ferry and Wolfe presented evidence for the presence of at least three strains in their consortium which degraded benzoate under anaerobic conditions.[67] One organism was able to degrade benzoate to intermediates such as acetate and formate. Two others were incapable of attacking benzoate but could produce methane from the intermediates. The removal of the intermediates was apparently necessary to provide thermodynamically-favourable conditions for benzoate degradation.

In another example Bryant et al. described a consortium of two members.[68] One oxidised ethanol to acetate and generated hydrogen. In monoculture of this organism, growth quickly ceased due to the accumulation of metabolic products. In the consortium the second organism removed the metabolic products and produced methane. A different type of interaction was described in a cyclohexane-metabolising consortium described by Stirling et al.[69] A species of Nocardia was responsible for cyclohexane degradation but could only grow in the presence of a species of Pseudomonas which provided it with several growth factors.

Most of the field trials of enhanced oil recovery conducted in Eastern Europe have employed consortia rather than individual strains. Lazar found that, in laboratory tests, the pure cultures and mixtures of pure cultures were much less effective at releasing oil than were bacterial consortia.[36] The consortia employed have not been fully described and no analysis has been made of the interactions occurring, but they have clearly been complex. Thus Jaranyi used an inoculum which was very rich in species including representatives of the genera Pseudomonas, Clostridium and Desulfovibrio while Karaskiewicz employed an inoculum containing Arthrobacter, Clostridium, Mycobacterium, Peptococcus and Pseudomonas.[70,71]

REFERENCES

1. Brock, T. D., Brock, K. M., Belly, R. T. and Weiss, R. L. (1972). *Arch. Mikrobiol.*, **84**, 54.
2. Heinen, W. (1971). *Arch. Mikrobiol.*, **76**, 2.
3. Jackson, T. J., Ramaley, R. F. and Meinschein, W. G. (1973). *Int. J. Syst. Bacteriol.*, **23**, 28.
4. Tansey, M. R. and Brock, T. D. (1978). In: *Microbial Life in Extreme Environments*, ed. D. J. Kushner, p. 159. Academic Press, New York.
5. Brock, T. D. and Darland, G. K. (1970). *Science*, **169**, 1316.
6. Bott, T. L. and Brock, T. D. (1969), *Science*, **164**, 1411.
7. Brock, T. D., Brock, M. L., Bott, T. L. and Edwards, M. R. (1971). *J. Bacteriol.*, **107**, 303.
8. Zeikus, J. G. (1979). *Enzyme Microb. Technol.*, **1**, 243.
9. ZoBell, C. E. and Kim, J. (1972). In: *The Effects of Pressure on Organisms*, eds. M. A. Sleigh and A. G. MacDonald, p. 125. Symposium No. XXVI, Society for Experimental Biology. Cambridge University Press.
10. Marquis, R. E. (1976). *Adv. Microbial Physiol.*, **14**, 159.
11. Marquis, R. E. and Matsumura, P. (1978). In: *Microbial Life in Extreme Environments* ed. D. J. Kushner, p. 105. Academic Press, New York.
12. ZoBell, C. E. and Budge, K. M. (1965). *Limnology and Oceanography*, **10**, 207.
13. ZoBell, C. E. (1970). In: *High Pressure Effects in Cellular Processes*, ed. A. M. Zimmerman, p. 85. Academic Press, New York.
14. Marquis, R. E. and Keller, D. M. (1975). *J. Bacteriol.*, **122**, 575.
15. Kriss, A. E. and Mitskevich, I. N. (1967). *Mikrobiologiya*, **36**, 203.
16. ZoBell, C. E. and Oppenheimer, C. H. (1950). *J. Bacteriol.*, **60**, 771.
17. Oppenheimer, C. H. and ZoBell, C. E. (1952). *J. Marine Res.*, **11**, 10.
18. ZoBell, C. E. and Cobet, A. B. (1962). *J. Bacteriol.*, **84**, 1228.
19. Boatman, E. S. (1967). *Bact. Proc.*, 25.
20. ZoBell, C. E. and Cobet, A. B. (1964). *J. Bacteriol.*, **87**, 710.
21. Maganathan, R. and Marquis, R. E. (1973). *Nature*, **246**, 525.
22. Marquis, R. E. and Bender, G. R. (1980). *Canad. J. Microbiol.*, **26**, 371.
23. Kushner, D. J. (1978). In: *Microbial Life in Extreme Environments*, ed. D. J. Kushner, p. 317. Academic Press, New York.
24. Dundas, I. E. D. (1977). *Adv. Microbial Physiol.*, **15**, 85.

25. Larsen, H. (1967). *Adv. Microbial Physiol.*, **1**, 97.
26. Forsyth, M. P., Shindler, D. B., Gochnauer, M. B. and Kushner, D. J. (1971). *Canad. J. Microbiol.*, **17**, 825.
27. ZoBell, C. E. and Johnson, D. M. (1949). *J. Bacteriol.*, **57**, 179.
28. ZoBell, C. E. (1958). *Producers' Monthly*, **22**, 12–29.
29. Albright, L. J. and Hanigman, J. F. (1971). *Canad. J. Microbiol.*, **17**, 1246.
30. Novitsky, T. J. and Kushner, D. J. (1975). *Canad. J. Microbiol.*, **21**, 107.
31. ZoBell, C. E. (1946). US 2,413,278.
32. Updegraff, D. M. and Wren, G. B. (1953). US 2,660,550.
33. Hitzman, D. O. (1962). US 3,032,472.
34. Coty, V. G. (1976). In: *The Role of Microorganisms in the Recovery of Oil. Proceedings of the 1976 Engineering Foundation Conference*, p. 77. National Science Foundation, Washington.
35. Dostalek, M. and Spurny, M. (1957). *Ceskoslov. mikrobiol.*, **2**, 300.
36. Lazar, I. (July 1978). *European Symposium on Enhanced Oil Recovery*, ed. J. Brown, Edinburgh, 5–7 VII 1978, p. 279. Heriot-Watt University, Edinburgh.
37. Jobson, A. Cook, F. D. and Westlake, D. W. S. (1972). *Appl. Microbiol.*, **23**, 1082.
38. Klug, M. J. and Markovetz, A. J. (1971). *Adv. Microbial Physiol.*, **5**, 1.
39. Swain, H. M., Somerville, H. J. and Cole, J. A. (1978). *J. Gen. Microbiol.*, **107**, 103.
40. Merkel, G. J., Underwood, W. H. and Perry, J. J. (1978). *FEMS Letters*, **3**, 81.
41. Davis, J. B. and Yarbrough, H. F. (1966). *Chemical Geol.*, **1**, 137.
42. Kuznetsov, S. I., Ivanof, M. V. and Lyalikova, N. N. (1963). *Introduction to Geological Microbiology* (English translation), ed. C. H. Oppenheimer. McGraw Hill, New York.
43. Karaskiewicz, J. (1968). *Nafta (Katowice)*, **24**, 198.
44. Abbott, B. J. and Gledhill, W. E. (1971). *Adv. Appl. Microbiol.*, **14**, 249.
45. La Riviere, J. W. M. (1955). *Antonie van Leeuwenhoek J. Microbiol. Serol.*, **21**, 1.
46. La Riviere, J. W. M. (1955). *Antonie van Leeuwenhoek J. Microbiol. Serol.*, **21**, 9.
47. Stewart, G. (July 1977) *Enhanced Oil Recovery*. Dept. of Petroleum Engineering, Heriot-Watt University, Edinburgh.

48. Gerson, D. F. and Zajic, J. E. (July 1979). *Process Biochem.*, 20.
49. ZoBell, C. E. (1953) US 2,641,566.
50. Martin, S. M. and Skerman, V. B. D. (1972) *World Directory of Collections of Cultures of Microorganisms.* Wiley-Interscience, New York & London.
51. Lindsay, J. A. and Creaser, E. H. (1975). *Nature*, **255**, 650.
52. Brock, T. D. and Boylen, K. L. (1973). *Appl. Microbiol.*, **25**, 72.
53. Kriss, A. E. (1962). In: *Marine Microbiology (Deep Sea)*, translated by J. M. Shewan and Z. Kabata, p. 91. Wiley Interscience, New York.
54. Kriss, A. E. and Zaichkin, E. I. (1971). *Mikrobiologiya* (English translation), **40**, 489.
55. Hopwood, D. A. (1970). In: *Methods in Microbiology*, eds. J. R. Norris and D. W. Ribbons, Vol. 3A, p. 363. Academic Press, London & New York.
56. Hayes, W. (1968). *The Genetics of Bacteria and Their Viruses*, 2nd edition. Blackwell, Oxford.
57. Larsen, H. (1962). In: *The Bacteria*, eds. I. C. Gunsalus and R. Y. Stanier, Vol. 4, p. 297. Academic Press, New York & London.
58. Brown, A. D. (1976). *Bacteriol. Rev.*, **40**, 803.
59. Pope, D. H. and Ogrinc, M. (1975). *Abstracts Ann. Meeting American Soc. Microbiol.*, p. 19.
60. Dowben, R. M. and Weidenmüller, R. (1968). *Biochim. Biophys. Acta*, **158**, 255.
61. Tempest, D. W. (1970). In: *Methods in Microbiology*, eds. J. R. Norris and D. W. Ribbons, Vol. 2, p. 259. Academic Press, London & New York.
62. Pirt, S. J. (1975). *Principles of Microbe and Cell Cultivation.* Blackwell, Oxford.
63. Clarke, P. H. (1974). In: *Evolution in the Microbial World*, eds. M. J. Carlile and J. J. Skehel, p. 183. *24th Symposium, Society for General Microbiology.* Cambridge University Press.
64. Friedman, S. M. and Mojica-A, T. (1978). In: *Biochemistry of Thermophily*, ed. S. M. Friedman, p. 117. Academic Press, New York.
65. Atherton, K. T., Byrom, D. and Dart, E. C. (1979). In: *Microbial Technology, Current State, Future Prospects*, eds. A. T. Bull, D. C. Ellwood and C. Ratledge, p. 379. *29th Symposium, Society for General Microbiology*, Cambridge University Press.
66. Slater, J. H. and Somerville, H. J. (1979). In: *Microbial Technology, Current State, Future Prospects*, eds. A. T. Bull, D. C. Ellwood and

C. Ratledge, p. 221. *29th Symposium, Society for General Microbiology.* Cambridge University Press.
67. Ferry, J. G. and Wolfe, R. S. (1976). *Arch. Mikrobiol.,* **107**, 33.
68. Bryant, M. P., Wolin, E. A., Wolin, M. J. and Wolfe, R. S. (1967). *Arch. Mikrobiol.,* **59**, 20.
69. Stirling, L. A., Watkinson, R. J. and Higgins, I. J. (1977). *J. Gen. Microbiol.,* **99**, 119.
70. Jaranyi, I. (1968). *M..All. Földtani Intezet Evi Jelentese Az 1968 Evröl,* 423.
71. Karaskiewicz, J. (1974). *Prace Instytut Naftowego, 3. Katowice: Wydawnictwo 'Slask'.*

CHAPTER 6

Getting Started: Matters of Logistics

In this chapter we shall consider some organisational problems which we see arising as part of the development of bacterial methods for the enhancement of oil recovery. For the most part we will have to pose questions rather than provide answers. There are two reasons for this: firstly, that there are so many contexts in which this work might progress, many perhaps requiring individual aspects of their treatment, that even if we knew all the answers we could not hope to deal with them in specific detail; and secondly, that the answers to some of the questions cannot at the present time be provided because of the lack of information. It is hoped, nevertheless, that by the presentation of the problem areas themselves lines of thought will be stimulated in the mind of the reader leading to solutions relevant for his own particular versions of the questions.

Furthermore, there will exist two strands of thought which it is impossible to keep entirely separate and which will of necessity be intertwined. These follow from the two main microbiological approaches already discussed in detail in earlier chapters: the molasses-based technique appropriate only for relatively shallow formations with narrow well spacings, or worked on a huff-and-puff basis, and the proposals for constructing strains or consortia for use in the deep reservoirs. The latter could equally serve the shallow reservoirs and the possibility of two ways of working these reservoirs and the relationship between them requires some consideration.

6.1. ARE GENERAL SOLUTIONS FEASIBLE?

One factor which must influence the way one sets about seeking solutions is the universality of application of methods which may be evolved. This question can be posed at two levels: the generality of the concept itself, and the detailed properties of the final operating system. Each of these levels applies separately to the two general microbiological techniques.

Experience to date with treatments involving the use of molasses as nutrient in shallow reservoirs suggests that the overall method for obtaining adapted bacterial populations and injecting them together with nutrient medium into a well can be employed successfully, at least with many wells. Following the recommendations of those workers that have used this technique, the adaptation of the bacterial population can make use of the particular bed waters and crude oil from the well in question, at the appropriate temperature and with other physical and chemical conditions as close as possible to those obtaining in the reservoir. Since the adaptation of the microbial population is a fairly simple *ad hoc* procedure there appears to be no great difficulty or expense in preparing populations tailored to individual situations.

The proposed techniques for use in the deep reservoirs are more problematic; for the moment they are indeed no more than proposals. However, it does seem likely that the general method for constructing bacteria with the required properties will be broadly applicable to the whole field. To what extent individual variations will be necessary to cope with differences between reservoirs we do not yet know, nor can we readily predict how difficult and time-consuming it may be to incorporate these variations into the systems under construction. What does seem probable is that individual tailoring, if it should be proved essential, will become easier to achieve as experience is gained in the synthesis of the special strains or populations.

6.2. IDENTIFICATION OF TARGET WELLS AND RESERVOIRS

Another aspect of initiating a programme of bacterial oil recovery is to decide on a series of wells or reservoirs which are to be the targets of the treatments. Factors important here are going to include physical (temperature, depth, pressure, permeability and porosity of the stratum) and chemical (composition of the crude oil, bed waters, flood waters, rock) features of the formation itself, geographical considerations of where the well is located, operational aspects (onshore/offshore, ease of access), commercial and production prospects (productivity of well, life expectancy, past production, relative production of oil, water and gas, investment of capital) as well as political pressures (governmental interest or lack of it; political stability of the region in question, taxation). What sort of formations are to be treated and by which of the two approaches? How many wells, or how extensive an area, is to be included? If the exercise is successful, how great a return is to be expected both in crude oil and in monetary terms? Over how long a period will successful enhanced recovery operate, and what production rate is necessary to render the whole operation worth while?

On the expenditure side one must ask what level of investment is acceptable in view of the probable rewards? Does this investment include the basic scientific research and development or will that be costed separately, perhaps funded from other sources? What would be the financial losses if the treatment were (a) not effective, or (b) damaging? Is such a loss acceptable? What proportion of treated wells could show losses if balanced by how much success in other treated wells? Are bacteriological facilities for the growth of organisms and measurement of their numbers in the produced water and oil to be provided on site? What would be the cost of doing so, including the cost of skilled personnel? (And we must point out here that microorganisms are very much alive; their successful use on an industrial scale necessitates their continual handling with care and understanding. They cannot be used as a chemical out of a bottle but must periodically be monitored for changes in properties and the presence of contamination. Hence there is a need for skilled microbiologists.) Or would the treated site be so conveniently located as to permit transportation of materials to and from a central laboratory?

If old wells were to be used, would they be in a good physical state? If not, would it be worth while restoring them? If bacterial enhancement methods meant that the production life of the well was to be significantly extended, would the maintenance and operating costs remain commensurate with a higher oil production for long enough to make the whole exercise profitable? In the event of new wells being treated, would this be expected to increase the flow of oil at an early stage or prolong an initial high rate for longer than would be the case without treatment? How would enhancement be evaluated (a problem we have discussed in earlier chapters)?

What would be the cost of oil produced in this way compared with that produced from a totally new well? What is the future world price of oil likely to be, and how will this anticipated price affect a decision to undertake enhanced recovery procedures? How is the future burden of taxation likely to change, and how will it influence the desirability of enhancing recovery? Will governments actively promote enhanced recovery procedures, and, if so, will they do so by tax advantages, subsidies or coercion? Will nationally-owned companies be at an advantage or a disadvantage if they should be obliged to undertake enhancement of recovery while privately-owned or foreign companies do not?

6.3. WHO IS GOING TO DEVELOP BACTERIAL ENHANCEMENT METHODS?

The pattern of research and development in this field in the past has been via activity mostly in oil company laboratories in the United States, and in government institutes in Eastern Europe. Those governmental laboratories, being official organisations, were also able to collaborate directly with the nationally-owned oil companies. Most of the fundamental scientific advances have taken place in university departments, but little applied work has been undertaken there.

It is, of course, clear that oil company involvement must be an integral part of technical development. It is to the companies, whether privately or publicly owned, that the wells belong and no field trials, let alone production runs, can take place without their active collaboration. They possess the information about well and reservoir characteristics which would be essential for identifying suitable targets for treatment. Without their help no applied work is possible, yet it is the experience of several workers in microbial enhanced recovery that the companies are reluctant at present to embark upon a serious programme for the development of these methods. We suspect this results partly from the low absolute rates of enhanced recovery so far obtained by the Eastern European workers, and partly because the companies have not so far seen the possibility of running these procedures on (to them) a worthwhile scale. They would quite reasonably like to be convinced of the value of the methods before undertaking to apply them in practice; yet without their collaboration the necessary prior information cannot be acquired. This dilemma has been noted before. However, even if companies are not prepared to carry the main burden of the development work they may well be sufficiently interested to offer technical advice, information and assistance to enable workers funded from elsewhere to make progress to the point where they do become interested. At that time the companies may wish to become directly and commercially involved.

The newer proposals for constructing specified bacterial strains by genetic manipulation are still too recent to have made much impact in the enhanced oil recovery field. But we suspect that a similar dilemma will arise with them and that until their commercial value has been demonstrated, at least in field testing, the oil companies will not be willing to undertake the costs of developing them. The methods will certainly cost more to bring to the state of readiness for field trials than will the molasses-based schemes since there are many difficulties to be overcome with the former while the latter have already been in use. Development and field testing of such a methodology would require some tens of man-years of effort, would probably take at least a decade to perfect and must be expected to cost some millions of dollars.

Another factor hampering commercial development of bacterial methods may be the uncertainty of patent protection.[1,2] Recent patent legislation regarding the protection of microorganisms and the procedures for obtaining them has not been extensively tested in the courts. Under the conditions of using bacteria to inject oil wells it might prove rather difficult to ensure that no organisms are acquired by competitors and if adequate protection via patents could not be relied upon the companies would quite properly be reluctant to embark upon expensive development work. On the other hand, oil wells are valuable assets: one doubts whether a company would choose to inject an illicitly-obtained organism without first ensuring that it would do the right thing. It may well be cheaper and less risky in both the long and the short runs to obtain a licence for the use of an organism or a technique and to modify it or confirm its acceptability for its intended use before putting it to work.

A reasonable course of action might thus be for the basic work to be done in university laboratories under governmental financial sponsorship together with technical collaboration from the appropriate oil companies. A phased development programme could be formulated and as each stage was successfully completed confidence would gradually build up in the concept as a whole. In this way one might hope that by the time the scientific progress had reached a stage commensurate with field trials there would be sufficient interest and confidence on the part of operators to make such field trials possible.

6.4. HOW URGENT IS IT TO START?

The question of deciding whether or not it would be most advantageous to inoculate a well with bacteria early in its productive life has already been discussed at length, together with the logistic difficulties associated with performing the experiment. It is brought up again now because its resolution must inevitably influence the sense of urgency which is applied to developing the technology. If it should turn out to be the case that a well not treated early is a well much of whose potential is lost for ever, then clearly the sooner bacterial methods are ready for use the better because new wells are continually being brought into production. If, on the other hand, bacterial enhancement is only a terminal procedure the urgency, though still present, is less acute since wells which will be brought on stream within the next decade will still be operational 10 or 15 years later and susceptible to terminal treatment. It is also conceivable that wells no longer productive and already shut down might be reactivated in the future and then subjected to bacteriological enhancement procedures. At the present time there is insufficient information to allow a decision as to the best time for inoculation. It seems prudent that if the methods are to be considered seriously, development should be started as soon as possible so as to give the maximum variety of options for their ultimate deployment.

6.5. THE MOLASSES-BASED TECHNIQUE

Because of the use already made of this method in Czechoslovakia, Hungary, Poland and Romania, and the information about it already published, its application in other countries by new groups of workers would be relatively rapid. Enough is known so that new groups would have strong leads on how to develop the methods to suit their own local conditions. Accordingly, development times would be expected to be short, and development costs fairly low. The method commends itself for widespread application to comparatively shallow, declining wells except for one unknown consideration. As it has never been run to ultimate exhaustion we do not know the mean fraction of oil ultimately recoverable when this method is used. Consequently, we do not know whether it might prove desirable to employ specifically constructed bacteria at a later stage, always assuming (and we cannot yet know this) that the latter will mobilise a larger fraction of the original oil-in-place. (We have placed the two methods in this time sequence because we suppose the molasses-based method could be brought to readiness several years before the construction of specifically designed strains.) And, most importantly, we do not know whether the use of the molasses-based technique in a particular section of a reservoir might preclude the subsequent deployment of the designed strains owing to the prior presence in the formation of large numbers of other bacteria. If the two techniques used in sequence within one formation were not compatible a decision would have to be made whether to employ the molasses-based method in the near future, or to wait a decade or more for the more sophisticated system to be perfected.

6.6. METHODS USING SPECIFICALLY DESIGNED BACTERIA

In previous discussion we have shown why the molasses-based approach is unlikely to be suitable for reservoirs with wide well spacings. Those reservoirs are also likely to be ones which are deep, hot and under high pressure, and for which bacteria would need to be constructed to meet their specific conditions. When methods had been worked out for constructing bacteria in that way, and their handling and operating conditions elucidated, it would, of course, be possible for the same approach to be used for the shallower reservoirs. Indeed, some of the difficulties to be overcome, particularly those related to temperature and pressure, would be less severe and as a result it would presumably be easier to construct custom-built bacteria for shallow formations than for deep ones. None the less, many of the technical problems would be common to both approaches, and in our view it would still take a distinctly greater effort to develop such bacteria for use even in the shallow reservoirs than it would for the molasses-based method: one would therefore have to decide in the cases of particular wells whether to wait for the custom-built strains or to proceed earlier with those feeding on molasses.

In developing the constructed forms there are several areas in all of which progress would need to be made in order finally to produce a successful outcome. These areas might be expected to include some or all of the following:

1. characterisation of the ability of bacteria to penetrate rock strata, and how this might be influenced by rock permeability, porosity, chemical composition, percentage saturation with oil, chemical composition of the formation waters, and by the size, shape, growth characteristics, deformability and motility of the bacterial cells;

2. characterisation of biochemical mechanisms capable of anaerobic metabolism of at least some components of the crude oil; the investigation of suitable electron acceptors, since molecular oxygen would necessarily be excluded;

3. identification of products of bacterial metabolism likely to be of value in mobilising crude oil;

4. isolation of bacteria able simultaneously to withstand the temperatures, pressures and salinities of the chosen reservoirs;

5. the possible need to develop genetic transfer techniques to permit all the required biochemical properties to be combined within one strain; or, if this is not possible, the development of consortia of different strains, each one of them able to survive in the conditions of the reservoir, which between them would be able cooperatively to carry out all the required biochemical reactions;

6. studies on the stability of the systems as constructed to ensure that they will continue to function for the required period in the actual conditions of the reservoirs;

7. testing under real operating conditions to confirm successful transfer from the laboratory to field trials.

6.7. OBTAINING WELLS FOR FIELD TRIALS

We shall not discuss again the difficulty of obtaining wells for field trials early in their productive lives but point out here that if information from such trials is essential then the difficulty must be overcome.

It need hardly be said that all trials involve costs and, being indeed trials, may yield no reward. If recompense is eventually obtained in the form of more oil there will inevitably be a delay until this comes on stream, during which time some operating costs would continue. So even the use of exhausted wells for trial purposes may require a significant financial investment.

However, exhausted wells are exhausted only by conventional technologies and contain most of their original oil still in the formations. Indeed, this is the very oil which is sought in tertiary recovery procedures. There exist many wells around the world, of very many types and characteristics, which could serve as model systems for terminal enhanced recovery using bacteria. Their formations lie at all depths and include some deep ones (for instance, in Trinidad) which could serve as models for new deep wells, such as those offshore in the North Sea. In these deep exhausted wells both the survival of constructed bacteria, and their ability to release more oil when their injection is coupled to a suitable waterflood, could be tested under genuine reservoir conditions.

6.8. COMPATIBILITY BETWEEN BACTERIAL AND OTHER RECOVERY METHODS

We asked above whether it might be possible to use purpose-built bacterial strains following injection of the reservoir with bacteria metabolising molasses and we recognised our inability to provide an answer. Other compatibility problems might exist between either of the potential bacterial methods and the use of non-biological floods.

Because it is not yet known how effective bacteria might be in releasing oil from the reservoirs, nor, indeed, how effective non-biological methods might be, we recognise that there may be advantage in using bacterial enhancement methods either before or after chemical flooding. Without experimental data it is not possible to come to any firm conclusions, but in the absence of evidence to the contrary, there seems to be no immediate reason why the prior use of bacteria should preclude a subsequent chemical flood, or vice versa, except for the possible deleterious effect on bacteria of petroleum sulphonates used in surfactant flooding. However, if these were first largely washed from the system the small residual quantities might be tolerated. It may perhaps not be impossible even to design and construct bacteria which could resist the action of such powerful surfactants ...

6.9. RESOURCES REQUIRED FOR BACTERIAL INJECTION

The molasses-based technique as practised so far has used a single initial injection of bacteria together with a charge of molasses, mineral nutrients and formation water. No wellhead facilities would be necessary, other than pumps for injection, as the bacteria would be grown in a laboratory or industrial fermentation plant which could be quite remote from the injection site. No difficulties are to be expected in connection with survival of the bacteria during transportation from the growth vessels to the wellhead. Should a second bacterial injection be decided upon later, or a supplementation of the nutrient, these would be effected by simple repetitions of the initial procedure.

With the proposals for employing specifically constructed strains or consortia we have as yet no experience of their use. It is conceivable that here, too, a single initial injection would suffice, with periodic booster injections if these should prove desirable. However, it may prove advantageous, in order to achieve a more ubiquitous distribution of the microorganisms, to inject them continuously with the waterflood. This would permit much lower cell population densities in the injected fluids and that, in turn, would probably significantly reduce the likelihood of plugging. No organic nutrient would be injected with this system since it is based on the presumption that the bacteria would be able to metabolise at least part of the hydrocarbon of the crude oil itself. It is to be expected that the waterflood receiving a continuous injection of bacteria would require mineral supplementation, particularly for the elements nitrogen and phosphorus, as well as with an electron acceptor. Conceivably two of these components could be combined into one compound as, for example, if nitrate were to serve simultaneously as the electron acceptor and the source of nitrogen.

Since no details have yet been elaborated for any particular case it is not possible now to predict how much mineral supplementation per unit volume of injected water would be required, and hence what it would cost. The extent of supplementation would no doubt depend both on the properties of the actual bacterial system and on the pre-existing mineralisation of the flood water. Some rough guesses and extrapolations might be made about the density of bacterial cells in the injected fluid. A concentration of 10^3 cells/ml would assume an average linear distance of 1 mm between cells. The rate of sea-water injection from one of the BP Forties platforms in the North Sea is about 1.5×10^5 bbl/day, or about 8.75×10^5 litres/h. Usually under laboratory conditions of bacterial culture it is easy to produce yields in excess of 10^9 cells/ml. If we take 10^3 cells/ml as being a reasonable cell density for injection, the stock culture could be diluted 10^6-fold into the waterflood, so that 8.75×10^5 litres/h of sea water would require 875 ml/h of bacterial culture. A bacterial growth rate of one doubling/h is commonplace, so that the size of the stock culture itself would need to be only 1225 ml if the cells for injection were being produced in a continuous culture apparatus. This means that even for wellheads remote from a microbiological laboratory (such as those on an offshore platform), and with very restricted working space, one could readily envisage a small-scale automatic culture apparatus which, even with supplies for a few days' operation, would not be very large or very heavy. It could probably be made simple and reliable enough not to require constant supervision by skilled operatives.

It is thus reasonable to conclude that, development costs aside, the running expenses of a continuous injection programme would not be very great. As it is not yet possible to estimate how efficient such a system might be in recovering more oil one can hardly draw up a balance sheet; nevertheless, if even a few per cent more oil were to be produced the economic benefits versus additional costs of operation would look very attractive indeed.

6.10. THE LONGER TERM VIEW

We conclude with some tentative questions about the long-term consequences of any successful enhanced recovery procedure, microbiological or otherwise, which was able significantly to increase the then current production rate, or future net yield, of oil for no more than a marginal increase in price.

Our evaluation at the present time of the molasses-based method is that it has been tested extensively enough to suggest that the additional development costs and expense of injecting a well would be small, and that useful additional supplies of oil might well be obtained in many local situations. But because it does depend on the use of a specific foodstuff it seems improbable that it could ever be used on a scale so huge that the additional yield of oil could actually alter significantly the world supply picture. Nevertheless, the method could probably be brought into use in hitherto unexploited areas with a lead time of three to five years, depending on the nature of the target reservoirs and the availability and experience of skilled microbiologists. Each successive successful deployment enhances the probability of its ever more widespread use.

Potentially, methods based on organisms able to feed on the crude oil *in situ* have a much wider application in practical use but they are unlikely to be of proven success for at least a decade. Even if they have proved successful by then in a limited number of cases, it might well take a further 10 years before their widespread use became common. In what sort of world fuel economy might such increased quantities of oil appear at the turn of the century? Even if efforts to develop the systems were started immediately, those making policy decisions for the next 10 or 15 years cannot be expected to take serious account of methods about to start development and not by any means assured of success. Thus, whatever plans may be made to develop energy and feedstock sources for the post-crude-oil economy will probably be 10 years further on before a microbial method of wide application can be demonstrated as being effective. Will such plans and investment in new facilities by then be so well entrenched that the additional supplies of oil which may be forthcoming will be seen as redundant? We ask this question, not with any intention of attempting to answer it, but as one more example of the complexity of the interactions which will influence a decision of whether or not to go ahead.

REFERENCES

1. Anon (15 March 1977). *New Scientist*, **81**, 845.
2. Bannister, D. (May 1979). *Soc. Gen. Microbiol. Quarterly*, **6** (3), 105.

CHAPTER 7

New Initiatives and Current Activities

While the work in Romania, reported in section 4.5.5, continues along previous lines, a number of new activities have begun to develop, particularly in the United States and in Britain. These are at a very early stage of development but may nevertheless represent a resurgence of interest in *in situ* microbial methods for the enhancement of oil recovery prompted, as we have suggested earlier in these discussions, by a combination of the increasing value of the hitherto unrecoverable oil and by new possibilities in the biological field.

7.1. UNITED STATES

7.1.1. Department of Energy

One of the first visible signs of renewed interest in the United States was a conference of interested parties sponsored by the US Department of Energy, and held in San Diego in the late summer of 1979.[1] After a series of brief general presentations the work of the conference was divided into four workshops: bioengineering (i.e. isolation and construction of microbial strains); reservoir ecology and environment; transformations (i.e. involvement of microorganisms in oil mobilisation directly in the reservoir); bioproducts (i.e. the properties and prospects for enhanced oil recovery of materials produced microbiologically in conventional fermentations and subsequently injected into the reservoir). The conference report lists in some detail a number of recommendations for action, applicable primarily to oil fields in the United States. The primary opportunities in that country for all types of enhanced oil recovery procedures lie in the very many (estimated by some to be in the tens, if not hundreds, of thousands) formally exhausted or low-producing wells distributed through a number of states. These are seen in the first instance as targets for the use of molasses-based or other nutrient-based microbial action. Many are relatively shallow, in fields with narrow well spacings, and may present no great environmental problems for microbial growth.

Presumably as a consequence of that meeting and of continuing interest by the Department of Energy a number of research contracts have now been placed, the objectives of which we summarise as follows:[2]

7.1.1.1.   *Isolation and Screening of Anaerobic Clostridia for Useful Characteristics*

Grula at Oklahoma State University is proposing to select microorganisms for use in enhanced oil recovery, paying particular attention to their quantitative and qualitative production of gases (carbon dioxide and methane), acids (acetic), solvents (alcohols and acetone) and surfactants. Mixtures of bacteria will also be used to test their ability to grow together in the conditions of shallow oil wells. It is intended to determine which solvent mixtures capable of being produced by bacteria may have the best action in mobilising oil and, having determined whether an optimum sequence for bacterial inoculation exists, to introduce it in field trials. Preliminary results are reported in reference 2 (pp. 161–2).

7.1.1.2.   *Biodegradation of Materials Used in Enhanced Oil Recovery*

In another study, Grula and Grula will determine the biodegradability and toxicity of a number of substances with a particular view to the consequences of spillage or accidental discharge to the environment. Their studies will, however, also be of immediate interest to those concerned with the possibility of such substances acting as substrates for undesirable bacterial growth in the reservoir strata themselves. One significant finding in this respect (reference 2, p. 163) is that the biodegradation of xanthan gum can occur anaerobically. A latex material (7703–77) was highly degradable when present as a linear polymer, and so was a substance identified only as an α-olefin sulphonate, although it was toxic to soil bacteria. Three other compounds, described as linear alkyl benzene sulphonates, were toxic and not biodegradable.

7.1.1.3. *Use of Microorganisms in Enhanced Recovery Processes*

At the University of Oklahoma, Bennett Clark will determine whether microbial cells can be used to plug selectively high-permeability zones and so improve sweep efficiency. As in other studies, he proposes to assess the value of microorganisms for enhanced oil recovery by their production of surfactants and biopolymers. Organisms capable of growing in oil reservoirs will be isolated (but the specific reservoir conditions are not specified) and their biochemical properties and potential roles in the recovery process evaluated. An interesting objective is to examine ways of oxygenating the oil reservoir which would, of course, greatly increase the range of options for *in situ* reactions, but also greatly increase the chances of stimulating the growth of undesirable organisms. These studies will be coupled with an investigation of the behaviour of microorganisms in oil-bearing formations, and together they will eventually lead to laboratory core tests and field trials (reference 2, p. 164). No experimental results are yet available.

7.1.1.4. In situ *Methods of Biopolymer and Biosurfactant Formation*

This is another project at the University of Oklahoma by Munnecke and Bennett Clark, with the objective of isolating organisms capable of growth above 55 °C in 5% sodium chloride on sucrose or crude oil as energy sources. Mechanisms will be determined for the activation of microbial metabolism *in situ* to generate sufficient quantities of polymer or surfactant to assist oil recovery, and a survey will be made using data files to define suitable target reservoirs. Some results are reported (reference 2, pp. 165–6): many organisms isolated from a variety of sources can indeed grow at 55 °C in 5% salt solutions, though the number is severely restricted if anaerobic conditions are imposed. High salt appears to be more restrictive than high temperatures. Most of the organisms are able to emulsify crude oil components suggesting the production of surfactant. The authors also reported low levels of biopolymer production, influenced in some cases by the carbon : nitrogen ratio in the medium.

7.1.1.5. Development of a Procedure for the Microbiological Evaluation of a Reservoir

Using cellular adenosine triphosphate as an indicator, Kujawa of Science Applications Inc., McLean VA proposes to define the biological activity of a petroleum reservoir (reference 2, p. 164). The proposal notes a lack at present of much of the information which would be needed to develop a successful microbial procedure for the enhancement of oil recovery. It observes that the petroleum reservoir–microbiological system must be known before a recovery technology can be developed, but it is not clear to the present authors what exactly is intended in this proposal nor how it would help to fill the gaps in current knowledge.

7.1.2. Private Sector

We are aware of only one non-governmental sponsored activity in the field in the United States. This was reported by Johnson at the San Diego meeting (reference 1, p. 30). The techniques he used are exactly in the stream of development which proceeded from ZoBell through various groups in the United States and Eastern Europe, and is still being used by Lazar in Romania (see Chapter 4).

Johnson uses species of *Bacillus* and *Clostridium* injected together with minerals and a nutrient source, often cattle feed molasses, but other organic substrates are sometimes used if molasses is unavailable. Generally copious quantities of gas are produced (carbon dioxide, methane and some nitrogen) which repressurises the well, reduces the viscosity of the oil and causes it to swell. Large amounts of acetic and other organic acids react with reservoir carbonates to form additional carbon dioxide, and help to clean up formation damage near the well bore. Surfactants, acetone and various alcohols are also released by the bacteria.

Under good conditions, and if the proper types of oil and reservoir conditions are chosen, an average increase in oil production in excess of 350% may be expected; the cost per additional barrel of oil is in the range 15–50 ¢. Low viscosity oils are best for treatment, with a gravity 15–30° API and a viscosity less than 400 cP. Carbonate reservoirs or those with carbonate cementation respond best because of their susceptibility to attack by the organic acids produced in the fermentation. High salt concentrations may diminish activity and the optimum reservoir temperature was rather low, at 40 °C, but the bacteria are reported to grow vigorously within a rather wide range (undefined) of reservoir temperatures.

Johnson's activities were conducted as practical procedures to improve recovery, and not as well-controlled scientific investigations. His own professional background is that of geology, not microbiology. It is therefore of no little interest that he has successfully employed these microbiological methods on an empirical basis in commercially successful operations.

7.2.   UNITED KINGDOM

Britain is the only country, apart from the United States, in which we are aware of recent new initiatives in the deployment of microbiology *in situ* for enhanced oil recovery. Her problems are very different from those of first importance in the American context. Although the United Kingdom possesses a small onshore oil production industry, and a programme of further onshore exploration is under way, the offshore industry is of overwhelming importance. The problems of using microorganisms in the deep, offshore fields are very great, as we have discussed in detail in Chapter 5.

A new research activity to explore some of the associated problems has come into existence at Queen Mary College, University of London, under the joint direction of the present authors acting together with Dr J.P. Robinson, and with sponsorship from the UK Department of Energy and the Science Research Council. The objectives in the first instance are fourfold:

1. to determine the possibility of bacterial growth on crude oil as a sole carbon and energy source in the complete absence of molecular oxygen;
2. to define the limits to bacterial growth as far as temperature, pressure and salinity are concerned, with a view to evaluating the likelihood of obtaining growth by any organism in the deep reservoirs;
3. to measure the propensity of bacteria to plug porous matrices, with the further intention of defining those properties of reservoir rocks and fluids, and of the different bacteria, together conducive to the penetration of the formation with a minimum risk of plugging;
4. to investigate the expression of mesophilic characters in thermophiles (see section 5.2.5 for significance).

No formal publications from this group have yet appeared but it has been reported that potentially useful bacteria have been obtained from aqueous or sediment samples exposed to oil or oil fractions, and that bacterial growth on crude oil could be initiated and supported in the absence of both oxygen and additional nutrients.[3] In work at high temperatures it has been found that continuous bacterial growth could be maintained in a turbidostat at 85 °C using a recognised thermophile and conventional nutrition.

7.3. CONCLUSION

The last two years have indeed shown that quickening of interest we have been expecting. It is likely that basic investigations directed towards using microorganisms in the reservoirs will increase in number and complexity for the next few years, by which time a sufficient body of knowledge will have accumulated to allow more serious commercial evaluation of possible practical procedures. If that stage is overcome successfully one may anticipate a very rapid expansion of research and development, one in which the operating companies will play a prominent role, leading ultimately to extensive field testing and perhaps the availability of a proven technology for production purposes. That is still a long way off but not quite so far, perhaps, as it appeared to be only a couple of years ago.

REFERENCES

1. *Conference on Microbiological Processes Useful in Enhanced Oil Recovery.* 29 August–1 September 1979 (published October 1979). Sponsored by the US Department of Energy (CONF—79 0871, UC—92).
2. Contracts for Field Projects and Supporting Research on Enhanced Oil Recovery (February 1981), *Progress Review* No. 24 (DOE/ BETC-80/4, UC-92a). US Department of Energy, Bartlesville Energy Technology Center, Bartlesville, OK.
3. Hawes, R. I. (February 1981). *Offshore Research Focus* (Department of Energy), No. 23, p. 6.

Glossary of Biological Terms

Aerobic: In the presence of oxygen; applied to microorganisms able or required to grow in the presence of oxygen.

Agar: Complex polysaccharide from seaweed used as an aqueous gelling agent.

Alga (*pl. algae*): A member of one of the lower orders of green plants, mostly microscopic in size, but including seaweeds.

Amino acid: Type of compound forming a monomeric unit of protein structure.

Anaerobic: In the absence of oxygen; applied to microorganisms able or required to grow in the absence of oxygen.

Autonomous: Semi-independent; applied to the ability of certain types of bacterial DNA to replicate independently of the chromosome.

Autoradiography: A technique of producing an image of a radioactive specimen by allowing the emitted radiation to impinge on a photographic emulsion.

Bacillus (*pl. bacilli*): A rod-shaped bacterium.

Bacterium (*pl. bacteria*): Type of microorganism, usually consisting of a single cell.

Base pair: A pair of nucleotides in nucleic acid which interact specifically via hydrogen bonding.

Biochemistry: Study of chemical processes and structures in living organisms.

Biogenic: Of biological origin.

Cell: Basic structural and functional unit of all living organisms except viruses.

Chromosome: A genetic linkage group; a structure bearing genes found in the nuclei of cells.

Clone: A population derived entirely from one individual.

Coccus (*pl. cocci*): A spherical bacterium.

Coenzyme: Non-protein cofactor required in many enzyme reactions.

Colony: Dense mass of cells produced from a single individual when microorganisms are cultured on solid media.

Conjugation: A sexual union of male and female bacteria for the exchange of genetic material.

Consortium: A population of bacteria of two or more types acting cooperatively and dependent on one another for growth.

Constitutive: An enzyme synthesised under all conditions of growth and not requiring a specific signal for synthesis to start or stop.

Contaminant: A microorganism whose presence in a particular situation is unintended and undesired.

Culture: A population of microorganisms, 'pure' if consisting only of one type and 'mixed' if two or more types are present.

Cytoplasm: Contents of living cells excluding the nucleus.

Deoxyribo(se)nucleic acid: Type of macromolecule, comprising genes and the bearer of inherited information; based on a polynucleotide chain.

Desulphurising bacteria: Organisms capable of reducing sulphur or its compounds.

Differential centrifugation: Technique of separating particles of different sizes and densities by subjecting them to different gravitation forces for varying periods in a centrifuge.

DNA: Common abbreviation for deoxyribonucleic acid.

DNA ligase: Enzyme catalysing the linking together, under certain conditions, of discrete pieces of DNA.

Electron acceptor: Substance capable of being reduced, thereby allowing another substance to be oxidised in a coupled reaction.

Endonuclease: Enzyme catalysing the hydrolytic splitting of specific chemical bonds within the bulk of a nucleic acid molecule.

Enteric: Within the intestine.

Environment: For any living cell or living organism, the whole of the universe outside its own boundaries.

Enzyme: Biological catalyst, always a protein.

Evolution: A continuous process of biological change resulting from chance mutation and subsequent selection.

Excretion: The elimination of the waste products of metabolism.

'Expression' of genes: The process of actually synthesising specific proteins on the basis of inherited genetic information.

Extracellular: Outside cells.

Feedback inhibition: The inhibition of the catalytic activity of one or more enzymes in a biosynthetic pathway by the end product of that pathway.

Female: The recipient in the sexual act of genetic exchange.

Fermentation: Incomplete biological oxidation of organic substrates in the absence of oxygen.

Flagellum (pl. flagella): A long, fine, hairlike structure projecting from a cell and whose beat enables the cell to swim.

Flora: Collective term for plants; also sometimes applied to bacteria, once formally categorised as primitive plants.

Free-living: Applied to an organism capable of living outside other organisms as distinct from those which can survive only within another.

Fungus (pl. fungi) = Mould: Member of a category of microorganisms usually filamentous in form though some, such as yeasts grow as single cells.

Gene: Unit of inheritance; in molecular terms a specific section of DNA carrying the information for either
1. the structure of particular protein, or
2. controlling the synthesis of particular proteins.

Genetics: Study of heredity and variation.

Genotype: The genetic (i.e. inherited) informational complement of an organism.

Genus (pl. genera): A term used in biological classification for organisms sharing many common characteristics and related to one another but not so closely as to be able to interbreed.

Germination: Onset of growth of a seed or spore, usually after a period of dormancy.

Growth: Increase in bulk of an organism, or increase in numbers of a population.

Inducer: Substrate specifically promoting the synthesis of a particular enzyme.

Inducible: Capable of being promoted by an inducer.

Inheritance: Information passed from parent(s) to offspring and required for the proper development of the latter.

Inhibition: Prevention of an activity, or a reduction of its rate or intensity.

Inoculum: Small number of microorganisms added to a culture medium from which a bulk quantity will ultimately develop.

Lysis: Degradative deterioration of cells resulting in disintegration and dissolution.

Male: The donor in the sexual act of genetic exchange.

Medium: Liquid surrounding a cell, or solidified gel on which it is resting, used commonly in microbiology.

Membrane: Thin structure composed of fats and proteins acting as a protective layer enclosing cytoplasm or separating it into different compartments.

Metabolism: The total series of biochemical reactions and interactions undergone by molecules in living cells.

Methanogenic: Applied to bacteria capable of generating methane.

Methylation: Modification of a molecule by adding methyl groups.

Microbe: Synonym for microorganism.

Microbial: Pertaining to microbes.

Microbiological: Pertaining to microorganisms or to microbiology.

Microbiology: Study of microorganisms.

Microorganism: An organism belonging to the categories of viruses, mycoplasmas, bacteria, fungi, algae or protozoa. Most of them are small, and the group owes its cohesion to the similarity of the techniques used to handle the various types.

Modification methylase: Enzyme possessed by microorganisms, etc., catalysing the methylation of certain bases of DNA in a pattern characteristic for each species.

Molasses: Unrefined sugar preparation.

Motile: Capable of self-generated movement.

Mutagen: Chemical substance or physical agent promoting the occurrence of mutations.

Mutagenesis: The various processes resulting in mutation.

Mutant: Organism or gene bearing a mutation.

Mutation: Inheritable change in the structure of a gene.

Nucleotide: Compound consisting of a pentose sugar, phosphoric acid and a nitrogenous base: a monomeric unit of nucleic acids such as DNA.

Nutrient: A substance required for nutrition; a food.

Permease: A membrane-bound enzyme catalysing the uptake of a specific substance from the medium into a cell, usually against a concentration gradient.

Phenotype: Collective physical characteristics of an organism determined by the interaction between its genotype and its environment.

Physiology: Study of function of organisms and their parts.

Plasmid: Genetic element, sometimes present in bacteria, never integrated into the bacterial chromosome, which replicates autonomously.

Polynucleotide: Linear polymer of nucleotides, basis of nucleic acids such as DNA.

Polypeptide: Linear polymer of amino acids, basis of proteins.

Protein: Molecule containing one or more polypeptide chains in which the order of amino acid residues in each chain is defined and usually invariant for any particular protein.

Protozoon (pl. protozoa): A member of one of the lower orders of animals, usually microscopic in size.

Regulation: Specific control of the occurrence or rate of a biochemical reaction.

Replication: Process by which two molecules of DNA are formed from each parent molecule in such a way that each of the daughter molecules is identical to the parent.

Repressible: Capable of being prevented by a repressor.

Repressor: Substance specifically preventing the synthesis of a particular protein.

Residue: That part of a monomeric unit, such as an amino acid or nucleotide, which remains when it is incorporated into a polymer, such as a polypeptide or a polynucleotide.

Restriction endonuclease: Enzyme catalysing the breakdown of the poly-nucleotide chain of foreign DNA in a manner characteristic of the species in which it occurs.

Retroinhibition: Synonym for feedback inhibition.

Screening: The technique of isolating desired microorganisms from many undesired ones by allowing colonies to form as clones from each individual, and testing each clone separately for the property under investigation.

Secretion: Discharge or extrusion of molecules from within the cell into the surrounding medium.

Selection: The phenomenon of the relative abilities of different organisms, in competition with one another, to survive in a certain environment as a function of their phenotype and hence of their genotype. In microbiology, the technique of isolating desired forms from many undesired ones by encouraging growth in an environment which prevents development of the undesired forms and permits multiplication only of the sought-after strain.

Sex: A type of genetic exchange between a male donor and a female recipient.

Species: A unit of classification; ideally a group of organisms so closely related that they interbreed to produce fertile offspring.

Spore: Small, reproductive form of a microorganism typically very resistant to unfavourable conditions.

Sporogenic: Capable of producing spores.

Stationary: A phase of microbial growth when all available nutrients are exhausted and the cells are in a relatively inactive metabolic state.

Sterile: Devoid of the power of reproduction. In microbiology commonly taken to mean devoid of living organisms. (N.B. The difference between these two usages is not real: reproduction is one of the innate properties of living matter. However, a microbial colony, unable to grow for whatever reason, may still be able to effect chemical change, and in that sense not be entirely 'dead'.)

Strain: A population of identical microorganisms; essentially synonymous with clone.

Strand: Of DNA, one of the two polynucleotide chains making up the molecule.

Substrate: In microbiology and biochemistry, molecules whose interaction is catalysed by a particular enzyme; also the nutrients supporting growth of a microorganism, especially organic compounds.

Sulphate-reducing bacteria: Those capable of reducing oxidised compounds of sulphur (such as sulphate), often with the production of hydrogen sulphide.

Transformation: Among bacteria a form of genetic transfer in which DNA is purified from the donor and is absorbed from the medium by the recipient.

Variant: Member of a population showing some differences in genotype from the norm, but not enough to belong to another species.

Variety: Unit of classification which is a subdivision of a species.

Vegetative cells: Those not engaged upon spore formation or sexual reproduction.

Viable: Capable of reproduction; capable of living. (N.B. These properties are indistinguishable; see note under Sterile.)

Virus: One of a category of minute microorganisms, unable to multiply outside other living organisms, and devoid of normal cellular structure.

Wild-type: Strain of a particular organism as isolated from a natural source; distinguished from various mutant strains which may be derived from it.

Yeast: Type of fungus, usually growing as single cells, or as clumps or cells, but not as filaments.

Index